COMPUTATIONAL
IMAGE QUALITY

COMPUTATIONAL IMAGE QUALITY

Ruud Janssen

SPIE PRESS

A Publication of SPIE—The International Society for Optical Engineering
Bellingham, Washington USA

Library of Congress Cataloging-in-Publication Data

Janssen, Ruud.
    Computational image quality / by Ruud Janssen.
      p. cm. – (SPIE Press monograph ; PM101)
    Includes bibliographical references and index.
    ISBN 0-8194-4132-5
    1. Imaging systems–Image quality. I. Title. II. Series.
TK8315.J35   2001
621.36—dc21   2001020614

Published by

SPIE—The International Society for Optical Engineering
P.O. Box 10
Bellingham, Washington 98227-0010
Phone: 360/676-3290
Fax: 360/647-1445
Email: spie@spie.org
WWW: www.spie.org

Printed in the United States of America.

Acknowledgment

I am deeply indebted to Frans Blommaert, my former Ph.D. supervisor at the Institute for Perception Research (IPO). His years of experience in visual research and his passion for David Marr's computational approach have been a great source of inspiration to me. I could never have accomplished this if it were not for the many discussions we had about the implications of Marr's levels, the advantages and disadvantages of doing top–down and bottom–up research, and regularization as a solution to all known problems.

Contents

COMPUTATIONAL
IMAGE QUALITY

Chapter 1

Introduction

Over recent decades the role of images in the communication of information has grown steadily. Advances in technologies underlying the capture, transfer, storage, and display of images have created a situation in which the use of images as a means of communicating information has become technologically and economically feasible. More importantly, however, images are in many situations an extremely efficient means to communicate information, as may be witnessed by the proverb "a picture is worth a thousand words." Without a doubt, this has been the most prominent factor pushing the technological development.

Notwithstanding these technological advances, the current state of the art still requires that certain compromises be made in the design of algorithms and devices for capture, transfer, storage, and display of images. Examples of such compromises are temporal resolution versus noise sensitivity (for capture), spatial resolution versus image size (for transfer and storage), and luminance range versus gamut (for displays). These and similar choices will, when made, affect the appearance of reproduced images, as well as the impression of how well the images are reproduced. To make optimal choices, it is therefore necessary to have knowledge of how particular design choices translate into the appearance of images and subsequently into the impression of how well these images are reproduced. In a nutshell, this is the central question of all image quality research.

Current image quality research can be divided into two fundamentally different approaches. The first approach focuses on experimental evaluation

(Roufs 1993). A typical setup would include a small group of human subjects judging quality, and possibly some related attributes such as sharpness, contrast, or colorfulness, of a set of displayed images that are manipulated to simulate the effects of several different design choices. In this way, the influence of these design choices on image quality can be measured and, by means of interpolation, approximately optimal choices can be made. Two serious drawbacks of this approach are (1) it is extremely time consuming, as well as tiresome for the participating subjects; and (2) the obtained knowledge cannot easily be generalized, since all relations found between design choice and image quality are descriptive rather than based on an understanding. As a result, in a single series of experiments only a small fraction of the possible design space can be investigated, and in practice the experimental procedure must be repeated for almost every possible set of design choices.

The second approach tries to address these drawbacks by means of the development of models that describe the influences of several physical image characteristics on image quality, usually through a set of image attributes thought to determine image quality. When the influence of a set of design choices on physical image characteristics is known, these models can be used to predict image quality instead of having to measure image quality experimentally. Two different types of models can be distinguished here. Both types share the common characteristic that image quality is expressed in terms of the visibility of distortions, or artifacts, introduced in the image as a result of certain design choices. Examples of such distortions are flickering, blockiness, noisiness, or color shifts. A hypothetical version of the image devoid of any such visible distortions is regarded as the "original," that is, the version of the image assumed to be of highest quality; and the visibility of distortions in the image, referred to as impairment, is used as a measure of quality degradation. The fundamental difference between the two types of model lies in how this impairment is calculated.

In the first type of model, physiologically or psychophysically inspired models of early visual processing are used to calculate impairment from a difference between two versions of an image, for example the "input" or "unprocessed" version, which is substituted for the original, and the "output" or "processed" version of a certain device or image processing algorithm. A well-known example of this approach is the JND map (just-noticeable differences map) presented by Daly (1993). The two most im-

2

portant drawbacks of this approach are (1) it is unclear what exactly the "unprocessed" version of an image is; and (2) what in fact is calculated here is the visible difference between two versions of an image, not image quality itself, and it remains unclear how the translation from visible difference to image quality should be performed. Usually this translation is made by means of fitting certain model parameters to experimentally obtained quality judgments of human subjects.

The second type of model differs from the first type in the sense that it tries to estimate visibility of distortions directly from the image instead of using a visible difference with an original. In this type of model, visible distortions of an image, such as unsharpness or noisiness, are predicted by estimating relevant physical attributes of the image, such as blur and noise spread. Using psychophysically established relations, these estimated physical attributes are then translated to visibility of distortions in the image. Finally, image quality is expressed in terms of these visible distortions using some kind of combination rule. The work presented by Nijenhuis & Blommaert (1997) is a good example of this approach. The uncertain translation from visibility of distortions to image quality, which usually must be fitted to experimentally obtained data, remains an important drawback of this approach.

As may be concluded from the above description of the state of the art in image quality modeling, images are regarded primarily as two-dimensional signals. Similarly, early visual processing is regarded as signal processing, with image quality being determined by a set of characteristics of the resulting output signal. There are a few serious shortcomings to this view, the most important of which is the fact that the fundamental question "What *is* image quality?" is never asked, nor answered. To give an answer to this question, based on a thorough understanding of visuo-cognitive processing and the role of images therein, is the very aim of this book.

The approach we will pursue in this book distinguishes itself from the above approaches in four fundamental ways:

- We will regard images not as *signals* but instead as carriers of visual *information*. Since an image is the result of the optical imaging process, which maps physical scene properties onto a two-dimensional luminance distribution, it encodes important and useful information about the geometry of the scene and the properties of the objects located within this scene.

- We will regard visuo-cognitive processing not as *signal processing* but instead as *information processing*. Following Marr (1982) and Newell (1990), we will regard vision and cognition as the processes in which, first, physical object properties, or correlates of these, are reconstructed from a two-dimensional luminance distribution using a set of dedicated computational algorithms and, second, resulting descriptions are interpreted by comparing them with descriptions stored in memory.

- We will regard visuo-cognitive processing not as an *isolated process* but instead as an *essential stage* in human interaction with the environment. Descriptions of scene content, as produced by visuo-cognitive processing, are used as input to subsequent stages in the process of interaction with the environment, such as semantic processing and action. Hence, visuo-cognitive processing plays a vital role within the interaction process.

- We will regard the quality of an image not in terms of the visibility of *distortions* in this image but instead in terms of the *adequacy* of this image as input to the vision stage of the interaction process. The basic definition of quality we will use is formulated in terms of the degree to which imposed requirements are satisfied. When images are considered to be input to the interaction process, requirements must be imposed on these images to guarantee the successfulness of the interaction process. In this view, the degree to which an image satisfies these requirements determines the quality of this image.

The structure of this book is as follows. In Chapter 2 "Approaches to Image Quality," we start by introducing some of the most prominent approaches to image quality modeling and prediction. In this chapter, we will briefly discuss these approaches and the philosophies behind them, and conclude with a discussion of the similarities and dissimilarities with the approach presented in this book.

In Chapter 3 "Image Quality Semantics," we formulate an answer to the question "What is image quality?" based on the philosophy outlined in the above four points. In this chapter, we give a description of image quality in terms of two components: *usefulness*, that is, the precision of the internal representation of the image; and *naturalness*, that is, the degree of match between the internal representation of the image and representations stored in memory. The results of two series of experiments are used to demonstrate the validity of this concept.

Chapter 4 "Visual Metrics: Discriminative Power through Flexibility" focuses on the internal quantification of outside world attributes. Using a rather technical view on visual processing, we regard vision primarily as a process in which attributes of items in the outside world are measured and internally quantified with the aim to discriminate and/or identify these items. In Chapter 4 we show that the scale function of metrics optimized with respect to these tasks should be (partially) flexible. Furthermore, we show that such metrics exhibit properties resembling phenomena such as adaptation, crispening, and constancy.

A straightforward implementation of the image quality concept of Chapter 3 using the partially flexible metrics of Chapter 4 is the topic of Chapter 5 "Predicting the Usefulness and Naturalness of Color Reproductions." In Chapter 5 a measure for usefulness is developed, based on the overall discriminability of the items in the image. Furthermore, a measure for naturalness of the grass, skin, and sky areas of the image is developed, based on memory standards for grass, skin, and sky color. These memory standards are themselves constructed from the grass, skin, and sky areas of a large set of images.

As its title suggests, Chapter 6 "Image Quality Revisited" returns to the concept for image quality introduced in Chapter 3. Following a strict, top-down analysis the entire trajectory is completed, from the semantics of image quality down to the development of algorithms for the prediction of usefulness, naturalness, and image quality. Chapter 6 therefore contains a thorough and explicit description of image quality according to the four-point philosophy outlined above.

Chapter 2

Approaches to Image Quality

In this chapter we will introduce some of the most prominent approaches to image quality modeling and prediction, namely *statistical measures, measures using visual front-end models, measures incorporating modulation transfer functions, multidimensional impairment measures, task performance measures,* and *measures for color reproduction quality.* We will briefly discuss these approaches and the philosophies behind them, and conclude with a short discussion of the approach presented in this book.

2.1 Statistical measures

This class of image quality measures originates from signal processing applications, where in some situations the need arises to compare an original signal with a distorted version. For example, a transmitted signal has to be compared with a received version, such that conclusions can be drawn about the reliability of the transmission medium. Image quality measures of this type use no characteristics of human visual processing whatsoever. Instead, they compare the luminance patterns of the original and distorted images on a pixel-by-pixel basis, and subsequently calculate statistical measures from this comparison. They therefore can be characterized as statistical signal-distortion measures.

One measure is used especially widely: the *mean square error* (MSE). For an image I and its reproduced version I', the MSE is calculated as follows:

$$\text{MSE}(I, I') = \frac{1}{NM} \sum_{i=1}^{N} \sum_{j=1}^{M} [I'(i,j) - I(i,j)]^2, \tag{2.1}$$

where $I(i,j)$ and $I'(i,j)$ represent the luminances of the original and distorted images for the pixel located at row i and column j, and where N and M represent the number of rows and columns of the images.

As can be concluded from Eq. (2.1), the MSE is symmetric for I and I', that is, $\text{MSE}(I, I') = \text{MSE}(I', I)$. In other words, the references "original" and "distorted version" can be chosen arbitrarily without influencing the value of the MSE. Therefore, the MSE is a measure for image quality *difference* only, and it cannot be used to predict which of the two versions of the image is better. Usually a "common sense" approach is followed, in which the original or transmitted image is regarded as the better image and any nonzero MSE is regarded as a loss in image quality. This assumption may sometimes be incorrect, for example in situations where the difference between the original and distorted versions of an image is the result of image processing stages explicitly aimed at image quality improvement.

A second widely known statistical image quality measure is the *peak signal-to-noise ratio* (PSNR). The PSNR is in fact a normalization of the MSE and is obtained by dividing the square of the luminance range R of the display device by the MSE and expressing the result in decibels:

$$\text{PSNR} = 10 \log_{10} \frac{R^2}{\text{MSE}}. \tag{2.2}$$

The foremost advantages of the PSNR with respect to the MSE are, first, that it is a dimensionless and normalized quantity and, second, that it increases monotonically with the "lack of difference," or *image fidelity*, between the original and distorted versions of the image. However, since it is based on the MSE it does share the above-mentioned disadvantage of being symmetric.

MSE and PSNR are among the first widely used image quality measures. Remarkably, opinions about their usefulness for image quality prediction differ to date. Although MSE and PSNR are often considered in the image quality literature as being inadequate measures (Cosman,

Gray & Olshen 1994, Eckert & Bradley 1998, Lubin 1995, Nijenhuis & Blommaert 1997, Winkler 1999, Zetzsche & Hauske 1989), experimental results presented by Martens & Meesters (1999) suggest that the MSE, when applied to CIE 1976 lightness (L^*) images instead of plain luminance images, performs equally well as a highly sophisticated model such as the *Sarnoff Visual Discrimination Model* (VDM) (Lubin 1995). These results were obtained for the case of images degraded by noise and blur and for the case of JPEG-coded images. Considering the amount of knowledge about the human visual system put into models such as the VDM, and considering that the VDM is explicitly designed to predict the quality of JPEG-coded images, this evidence is surprising at least.

2.2   Measures using visual front-end models

The statistical measures from the previous section incorporate no knowledge of human visual processing whatsoever, which has been the primary reason for the development of visual front-end models for image quality prediction. The rationale for the development of this type of model is straightforward: the more characteristics of human visual processing are incorporated into image quality models, the better these models are likely to be at predicting the perceived difference between a reproduced image and its original, and hence the better these models should be able to predict image fidelity. This development has led to several highly complex, physiologically and psychophysically inspired models (Daly 1993, Lubin 1995, Zetzsche & Hauske 1989), of which the VDM and the *Visual Differences Predictor* (VDP) (Daly 1993) are the most widely known.

Visual front-end models for image quality prediction usually incorporate a wide range of physiologically and psychophysically established mechanisms of human visual processing. Typical components include

- optical blurring, to model the effect of the imperfect optics of the eye;

- luminance adaptation, to account for the variation in visual sensitivity as a function of light level;

- contrast sensitivity function, to account for the variation in visual sensitivity as a function of angular frequency;

- decomposition into multiple frequency bands, to model the spatial frequency selectivity of the human visual system;

- filtering by orientation-selective filters, to model the orientation selectivity of the human visual system;

- luminance and contrast masking, to account for the variation in visual sensitivity as a function of background structure.

Most models furthermore include a rather ad-hoc error summation stage to obtain a visible differences image from the scale-space descriptions of the original and distorted images generated at earlier stages. The visible differences image indicates, for each location in the image, the probability of seeing a difference between the original and distorted images. In some models the visible differences image is collapsed into a single number.

To conclude, it is important to note that front-end models are in fact threshold metrics; that is, they are designed to predict the visibility of distortions near the visual threshold level. When distortions are near the threshold level, their predictions are usually more consistent with observer ratings than the statistical measures. However, for suprathreshold distortion levels performance is often less good (Eckert & Bradley 1998), possibly due to the lack of modeling of higher-level visual processing.

2.3 Measures incorporating modulation transfer functions

This class of measures uses the modulation transfer function (MTF) of the display system in combination with the contrast sensitivity function (CSF) of the eye to predict the effects of, primarily, resolution and luminance contrast of a display device on image quality. Perhaps the best-known example of this type of measure is the *square-root integral* (SQRI) (Barten 1990).

The MTF of a display system expresses the modulation transfer of a displayed one-dimensional sinusoidal pattern as a function of spatial frequency. A typical MTF shows the characteristics of a low-pass filter, with a relatively constant modulation transfer for frequencies lower than the cutoff frequency, and rapidly decreasing modulation transfer for higher frequencies. Two properties of the MTF are especially important for the quality of displayed images. First, higher values of the modulation transfer lead to higher luminance contrast of the displayed pattern. Second, higher values of the cutoff frequency allow for higher-frequency patterns to be displayed. Both properties together generally result in higher contrast and sharper detail in displayed images, and therefore to higher image quality. An obvious way to characterize the quality of a display system therefore is the area under its MTF curve.

However, the sensitivity of the human eye varies with angular frequency and with average luminance level. The function expressing the contrast sensitivity of the human eye as a function of angular frequency is known as the *contrast sensitivity function* (CSF). In measures such as the SQRI, the CSF is therefore used as a weighting function to the MTF. Barten (1990) uses the following analytical expression to approximate the CSF:

$$\mathrm{CSF}(u) = aue^{-bu}\sqrt{1 + ce^{bu}}, \tag{2.3}$$

with a, b, and c given by

$$a = \frac{540(1 + 0.7/L)^{-0.2}}{1 + \frac{12}{w(1 + u/3)^2}}, \tag{2.4}$$

$$b = 0.3(1 + 100/L)^{0.15}, \tag{2.5}$$

$$c = 0.06, \tag{2.6}$$

where u is the angular frequency in cycles per degree, w is the angular size in degrees, and L is the average luminance in candelas per square meter. Since the CSF expresses contrast sensitivity as a function of angular frequency and the MTF expresses modulation transfer as a function of spatial frequency, a conversion must be made between spatial frequency and corresponding angular frequency. This can be done easily once the viewing distance is known.

Barten (1990) argues that nonlinear behavior of the human eye must be taken into account; first, by using the square root of the weighted MTF and, second, by using logarithmic integration. The SQRI is therefore calculated as

$$ \text{J} = \frac{1}{\log 2} \int_0^{u_{max}} \sqrt{\text{CSF}(u)\text{MTF}(u)} \left(\frac{du}{u} \right), \qquad (2.7) $$

where scaling by the factor $1/\log 2$ is done to yield results in units of 1 JND.

MTF methods like the SQRI are used primarily to predict the effects of display system parameters such as resolution, average luminance, luminance contrast, display size, and viewing distance. Barten (1990) reports good linear correlations between SQRI predictions and quality judgments by human observers for situations in which one or more of these parameters are systematically varied. However, since MTF measures only consider the display system and not the image itself, they cannot be used to predict image quality when factors other than those directly related to the display system are involved.

2.4   Multidimensional impairment measures

The quality of an image can be affected by a large variety of distortions. Some distortions can be one-to-one related to a visually distinguishable effect they have on the appearance of the image. For example, blurring of the two-dimensional image signal will lead to an unsharp image, and adding noise to the image signal will result in a noisy image. Other distortions can be related to more than one visually distinguishable effect. For example, blockiness and ringing are two visually distinguishable effects caused by JPEG compression of the image signal.

A visually distinguishable effect caused by a distortion of the image signal is usually referred to as an *impairment*, whereas the perceived magnitude of the effect is referred to as the *impairment strength*. The central assumption used in the multidimensional impairment approach is that image quality is inversely related to total impairment strength I_{tot} (Nijenhuis & Blommaert 1997); that is,

$$ Q = 1 - I_{tot}. \qquad (2.8) $$

Hence, when image quality is affected by more than one impairment, total impairment strength must be calculated from the individual impairment strengths. Usually, this is done by Minkowski summation of the individual impairment strengths (de Ridder 1992):

$$I_{tot}^{\alpha} = \sum_{i=1}^{N} I_i^{\alpha}, \qquad (2.9)$$

where I_i is an individual impairment's strength, N the number of impairments, and α the Minkowski exponent. For visually clearly distinguishable impairments, α is approximately two, whereas for impairments that are hardly distinguishable, α is approximately one. By using Minkowski summation, the location of a degraded image may be specified in a multidimensional impairment space of which the individual impairments are the dimensions, such that the distance (Euclidean for $\alpha = 2$ or City-block for $\alpha = 1$) of the image to the origin of the multidimensional space corresponds to total impairment strength.

An important aim of the multidimensional impairment approach is to find (psychophysical) relations between, on the one hand, physical parameters specifying the distortion of the image signal (for example, spread of a blurring kernel) and, on the other hand, the perceived strength of the resulting impairment (unsharpness). Typical examples of the multidimensional impairment approach in the literature are Kayargadde & Martens (1996a) and Kayargadde & Martens (1996b), who consider images degraded by noise and blur, and Nijenhuis & Blommaert (1997), who consider images degraded by sampling and interpolation artifacts. For example, the following relation can be used to predict unsharpness I_b from estimated blur kernel spread σ (Nijenhuis & Blommaert 1997):

$$I_b \propto 1 - \frac{1}{[(\sigma/\sigma_0)^2 + 1]^{0.25}}, \qquad (2.10)$$

where σ_0 represents the intrinsic blur due to optical and physiological properties of the human eye.

The multidimensional impairment approach has one clear advantage over "difference measures" like the statistical measures and the visual front-end measures: there is no need to consider the original version of an image. Impairment strength can be calculated directly from the image by estimating relevant physical parameters, such as blur kernel spread or noise variance, and by relating these parameters to perceptually relevant attributes, such as unsharpness or noisiness, using psychophysically established relations such as Eq. (2.10). Disadvantages of this approach are that it is still unclear whether a hypothetical "unimpaired" version of an image is indeed the version with the highest possible image quality. Furthermore, quite a large set of impairment measures may be necessary to adequately predict image quality in "real life" situations, where the image may be distorted by a complex ensemble of influences.

2.5 Task performance measures

The approach followed here differs significantly from the other approaches. Here, image quality is defined in terms of *task performance*. The image is considered to be of good quality when a well-defined task, such as detection of a certain lesion in an x ray scan, can be performed correctly. This approach is primarily used in three situations. First, it is used for medical applications, where diagnostic accuracy is a highly important issue. Second, it is used for military applications, where object detection and recognition are important for reconnaissance tasks. Last, it is used in office applications where readability of displayed text is important.

At the present, no adequate models exist that can reliably predict diagnostic accuracy, object detection and recognition, or text readability directly from the properties of the image. This is the main reason why task performance measures are usually derived during experiments. To this end, each field (medical, military, office) has developed its own methodology. We will discuss these methodologies briefly.

2.5.1 Diagnostic accuracy

In the medical field, the *receiver-operating characteristic* (ROC) is often used as a measure for diagnostic accuracy (Cosman et al. 1994). The ROC can be traced back to signal detection theory, where a detector must decide

14

Table 2.1: The four situations that can arise when a signal detector must decide whether a signal is present in background noise. The two decisions on the diagonal correspond to correct decisions, the other two correspond to incorrect decisions.

	Signal present	Signal absent
Signal detected	Hit	False alarm
Signal not detected	Miss	Correct rejection

whether or not a signal is present in background noise. When detection is performed by comparing the input to a fixed threshold, the four situations listed in Table 2.1 can arise.

An important way to characterize the detector performance is to systematically vary the detector threshold and to plot the hit rate versus the false alarm rate for several threshold settings. The resulting curve, which is known as the ROC curve, connects the points $(100\%, 100\%)$ (for an extremely low threshold) and $(0\%, 0\%)$ (extremely high threshold). The area under the ROC curve is used as a measure for detector performance.

In medical situations, the radiologist typically assumes the role of the detector, while the signal to be detected often is a certain type of lesion in an x ray or MRI scan. The quality of the medical image is characterized by the area under the experimentally measured ROC curve for the radiologist. The "detector threshold" of the radiologist is varied indirectly; for example, by systematic manipulation of a reward–penalty system on correct and incorrect decisions. Measuring the quality of medical images using ROC analysis therefore is a highly labor-intensive task.

2.5.2 Reconnaissance

In the military field, the *National Image Interpretability Rating Scale* (NIIRS) is well established (Hermiston & Booth 1999, Leachtenauer 2000). The NIIRS is a 10-level numerical scale. At each level, reconnaissance tasks ("criteria") are defined for target types such as vehicles, ships, and electronics. The higher the NIIRS level, the more demanding these criteria become with respect to the amount of information that must be extracted from the image. NIIRS ratings are given by certified image analysts. When rating an image, the image analyst indicates the highest NIIRS level that is satisfied by the image. In other words, the image analyst estimates the most difficult task that he or she believes can be performed using the image.

2.5.3   Text readability

In the office environment, *reading speed* and *reading comfort* are most often used to characterize the quality of displayed text. Roufs & Boschman (1997) and Boschman & Roufs (1997) give a comprehensive overview of methodological and experimental issues related to the measurement of readability of text displayed on a video display unit (VDU). Among the experimental variables they consider are ratings of reading comfort as reported by human subjects, reading performance as measured in terms of reciprocal search time in a letter-search task, and eye movements as measured in terms of fixation duration and saccade length. None of these measures shows a clear advantage over the others, which is why the authors conclude that ratings of reading comfort are probably sufficient

2.6   Measures for color reproduction quality

Basically all approaches mentioned in the preceding sections can be extended to predict the quality of color reproductions. For example, statistical measures such as *MSE* can be defined using the CIELAB or CIELUV ΔE color difference metrics recommended in 1976 by the Commission Internationale de l'Éclairage (CIE). These metrics use the difference between the CIELAB (L^*, a^*, b^*) or CIELUV (L^*, u^*, v^*) color coordinates of a pixel in the original and reproduced images to calculate a color difference ΔE using a simple Euclidean distance metric:

$$\Delta E_{ab} = \sqrt{(\Delta L^*)^2 + (\Delta a^*)^2 + (\Delta b^*)^2}, \tag{2.11}$$

$$\Delta E_{uv} = \sqrt{(\Delta L^*)^2 + (\Delta u^*)^2 + (\Delta v^*)^2}. \tag{2.12}$$

A well-known improvement on the CIELAB ΔE metric is the S-CIELAB ΔE metric proposed by Zhang, Setiawan & Wandell (1997) and Zhang & Wandell (1998). S-CIELAB ΔE is a spatial extension of CIELAB ΔE that aims to include the spatial-color sensitivity of human color vision. S-CIELAB ΔE values can be thought of as CIELAB ΔE values calculated for spatially filtered versions of the original and reproduced images.

A very different approach to modeling the quality of color reproductions, one which takes into account explicitly the influence of higher-level visual processing, is known as the *naturalness constraint* (de Ridder 1996). The naturalness constraint expresses the idea that an image of high quality should at least be perceived as realistic, or natural. For color reproductions this implies that an object's color as reproduced in an image should resemble the object's prototypical color as remembered from past observations in real life. Experimental data indeed support this idea: image quality judgments of human observers were found to be highly correlated with naturalness judgments (de Ridder 1996, Fedorovskaya, de Ridder & Blommaert 1997). In these and similar experiments, a second factor that was found to influence the quality of color reproductions is *colorfulness*, that is, the vividness of the colors in the reproduction. It was found that, in general, observers show a clear preference for more colorful, yet slightly unnatural reproductions (de Ridder 1996).

Yendrikhovskij, Blommaert & de Ridder (1999a) have taken this approach one step further by developing algorithms to predict naturalness and colorfulness from the color statistics of the image. In the proposed algorithms, colorfulness is estimated from statistical properties of the saturation of the colors in the image, whereas naturalness is estimated by comparing the color coordinates of the image areas representing grass, skin, or sky with experimentally measured prototypical color coordinates for grass, skin, or sky. Image quality can be obtained from colorfulness and naturalness estimates by means of a linear combination; that is,

$$Q = wN + (1 - w)C, \qquad (2.13)$$

where Q is image quality, w is a weighting factor between zero and one, and N and C are the estimates for naturalness and colorfulness, respectively.

2.7   The approach presented in this book

The approach presented in this book can be thought of as a theoretical underpinning of the approach followed by Yendrikhovskij et al. (1999a). To start, image quality is regarded as the degree to which the visual and cognitive systems are able to *use* the information presented in the image. Two criteria are derived for this: *discriminability* and *identifiability* of image content. Discriminability is related to the average difference between the colors in the image (and hence related to colorfulness), whereas identifiability is related to the similarity between object colors in the image and object colors in memory (and hence related to naturalness). Since image quality is defined here in terms of the performance of visuo-cognitive processes, this approach may also be thought of as a task-performance approach.

The aim of this approach is not just to model or predict image quality. Instead, the aim is to *understand what image quality is*. To this end, our default assumption is that visual processing of images is a goal-directed process. To successfully achieve the goal of visual processing, requirements must be imposed on the image. The logical consequence of this view is that the quality of the image is defined by the degree to which these requirements are satisfied.

Chapter 3

Image Quality Semantics

In this chapter we will discuss image quality in the context of the visuo-cognitive system as an information-processing system. To this end, we subdivide the information processing as performed by the visuo-cognitive system into three distinct processes: (1) the construction of a internal representation of the image; (2) the interpretation of this representation by means of a confrontation with memory; and (3) task-directed semantic processing of the interpreted scene in order to formulate a proper response.

A successful completion of these processes can be ensured only when two main requirements are satisfied: (1) the internal representation of the image is sufficiently precise; and (2) the degree of correspondence between the internal representation and "knowledge of reality" as stored in memory is high.

We then relate these requirements to the attributes "usefulness" and "naturalness" of the image, and give a functional description of image quality in terms of naturalness and usefulness. To conclude, experimental results supporting this description of image quality will be discussed.

This chapter is a slightly modified version of Janssen & Blommaert (1997), Image quality semantics, *Journal of Imaging Science and Technology*, vol. 41, no. 5, pp. 555–560. Reprinted with permission of IS&T, The Society for Imaging Science and Technology, Springfield, VA, USA.

3.1   Introduction

A major part of research activity in the field of image quality is directed toward the development of reliable, widely applicable, instrumental image quality measures. There are two important reasons for developing instrumental measures: (1) quality evaluation by means of subjective assessment tests is quite expensive and time consuming; and (2) a posteriori assessment of the image quality of a given design does not allow for an a priori optimization of this design, thus condemning the design of image (re)production systems to remain an iterative procedure.

At present, much of the research concerning instrumental image quality measures is based upon an approach that can be characterized as a "signal evaluation" approach. In this approach, the image is regarded as a complex signal that deviates more or less from the complex signal that represents the ideal or "original" image. Images are defined in the physical or perceptual domain, in the latter case using models of the earliest stages of visual perception, and quality measures are defined as distances in an appropriate function space, for example, Euclidean distance between actual and original image.

In contrast to this approach, we regard the processing of images by the visuo-cognitive system not as the evaluation of complex signals but instead as the *processing of visual information*. Realizing that this information processing is an essential part of an observer's interaction with his environment, we characterize the quality of an image in a more meaningful manner as the degree to which the image can be successfully exploited by the observer. We will therefore consider the visuo-cognitive system as (1) an information processing system (Marr 1982, Barrow & Tenenbaum 1986, Eimer 1990, Watt 1991); and (2) an integral part of an observer's interaction with his environment (Bruce & Green 1985, Gibson 1950, Gibson 1966, Gibson 1979).

3.2   Quality and information processing

3.2.1   Understanding information-processing systems

We adopt the general viewpoint in computational cognition (Newell & Simon 1972, Newell 1990) that (1) the visuo-cognitive system can be considered to be an information-processing system; and (2) information-processing systems can only be completely understood when they are understood at three distinct levels. These levels are (1) the semantic level; that is, the level describing the system in terms of computational goals and strategies; (2) the algorithmic level; that is, the level describing the implementation of the computational theory into algorithms and associated representations; and (3) the level of physical implementation; that is, the level describing the physical implementation of these algorithms and representations.[1]

As stated above, any information-processing system can only be completely understood when it can be appropriately described at *all three* levels, which for reasons of simplicity may be designated as the levels of "what and why," "how," and "where." The approach we choose to pursue here is strictly top-down; that is, we first try to gain a fundamental understanding of the "what and why" of the processing of images by the visuo-cognitive system in order to arrive at an understanding of what image quality is, and then proceed with the "how" and "where." Our present purposes are therefore served best with a description of visuo-cognitive information processing at the semantic level.

3.2.2   The quality of information

At a semantic level, the interaction of an observer with his environment can be described by a cycle consisting of three activities: (1) perception; that is, acquiring information from the environment and constructing an internal representation from it; (2) cognition; that is, interpreting the obtained internal representation; and (3) action; that is, responding appropriately to this interpretation.[2]

At this point, we may already infer that in order to ensure a proper response to occurrences in the outside world, certain requirements must be imposed upon the information that is acquired from the environment. When the

---

[1]Marr (1982) refers to these levels as the level of the *computational theory*, the level of *representation and algorithm*, and the level of the *hardware implementation*, respectively. The semantic level therefore corresponds to Marr's computational level.

[2]Throughout this book we will use the term cognition rather loosely to refer to almost the entire set of processes following perception and preceding action. Note, however, that action may depend on other inputs as well, such as emotions.

quality of information is considered to be the degree to which these requirements are satisfied, we arrive at two important conclusions: (1) the quality of information can be defined only within the context of an observer interacting with his or her environment; and (2) the quality of information refers to the appropriateness of this information as a basis for a proper response to outside world occurrences.

3.2.3 The quality of images

Realizing that images are the medium for visual information, we now focus our attention on the question of what requirements should be imposed upon an image. At first thought, the requirements that a "good" image should satisfy would seem to be precision and reliability. The above outlined ideas, however, lead to a somewhat more complicated requirement: for an image to be of "good" quality, the observer's *interpretation* of this image should be successful; that is, it should with high probability be correct. The imposition of this requirement is justified by realizing that *it is of vital importance that there is no discrepancy between "what really is there" and what the observer assumes is there.* The next section will focus on the question how this can be optimally secured.

3.3 Image quality semantics

3.3.1 Image processing by the visuo-cognitive system

We will subdivide the processing of images by the visuo-cognitive system in three distinct processes (Marr 1982, Barrow & Tenenbaum 1986): (1) perception, that is, the construction of a internal representation of the image using primarily low-level knowledge of the visual world; (2) interpretation, that is, the confrontation ("matching") of this internal representation with memory representations; and (3) task-directed semantic processing of the interpreted scene in order to formulate a response. A diagrammatical depiction of the processing of images by the visuo-cognitive system is shown in Fig. 3.1.

The collection of memory representations of the outside world, which we will from this point onward refer to as "knowledge of reality," we regard as

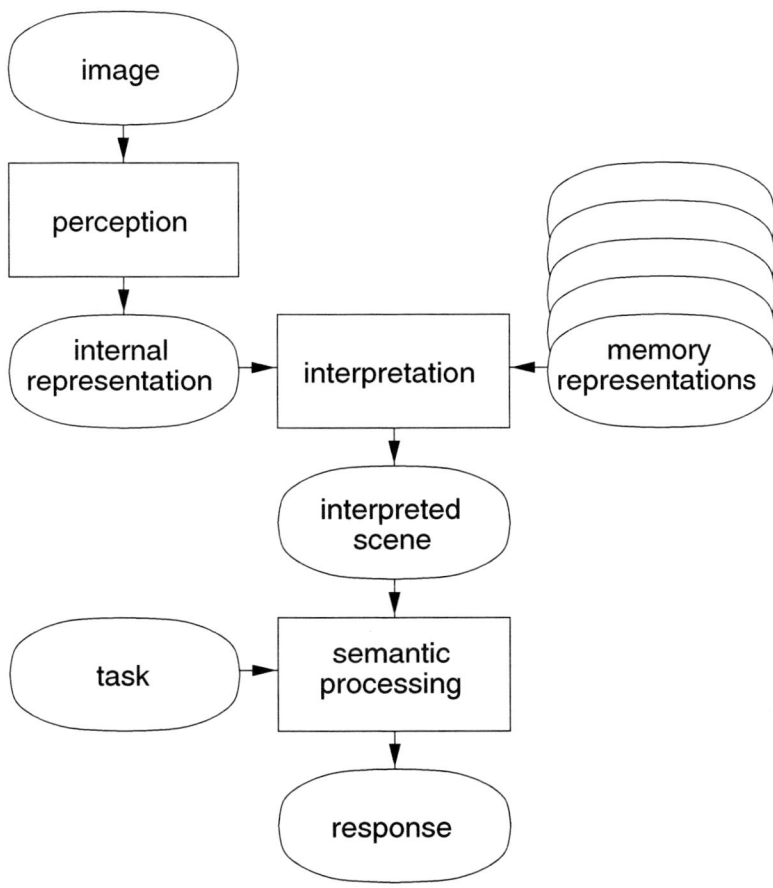

Figure 3.1: A diagrammatical depiction of visuo-cognitive processing of images. In this diagram, ellipses denote representations of information, and rectangles denote processes transforming one representation into another. Note that we consider the "response" to be a formulated sequence of actions, *not* its manifestation in terms of motoric events.

well defined but nevertheless fuzzy. As an example of this, consider that although observers "know" what the prototypical characteristics of a certain object are, they usually are unable to define a clear distinction between prototypical and nonprototypical object characteristics. Knowledge of reality therefore may be thought of as accumulated knowledge of (the behavior of) outside world statistics. Instead of referring to a match between the internal representation and a memory representation, we will henceforth refer more appropriately to a match between the internal representation and knowledge of reality.

3.3.2   Naturalness, usefulness, and quality

Given the visuo-cognitive processes as outlined above, we now ask how a successful interpretation of an image can be best secured. Returning to Fig. 3.1, we may readily conclude that for a successful interpretation of an image, the interpretation process should result in a satisfactory match between the internal representation and knowledge of reality. We therefore arrive quite directly at two principal requirements that an image of "good" quality should satisfy: (1) the internal representation of the image should be sufficiently precise; and (2) the degree of correspondence between the internal representation and knowledge of reality as stored in memory should be high.

We are now able to formalize the preceding discussion by defining (1) the *usefulness* of an image to be the precision[3] of the internal representation of the image; and (2) the *naturalness* of an image to be the degree of correspondence between the internal representation of the image and knowledge of reality as stored in memory. Using these definitions, we define the quality of an image to be the degree to which the image is both useful and natural.

The sets of requirements that one needs to impose upon an image in order to maximize the usefulness or the naturalness of this image will in general not coincide. For example, detection or discrimination of objects in an image may require "exaggeration" of certain features of this image, resulting in a less natural reproduction of the image. Therefore, given the above def-

---

[3]We use the term precision here to refer to a kind of internal signal-to-noise ratio. Precision therefore does not necessarily mean a one-to-one correspondence with reality.

inition of image quality, we postulate that the quality of an image will be given by a compromise resulting from simultaneously evaluating to what degree the image satisfies the sets of requirements that lead to maximizing the usefulness or the naturalness of the image.

3.4 Experiments

3.4.1 Experiment 1: Influences of naturalness and usefulness on image quality

Aim

In order to test the semantic description of image quality as given above, we performed an experiment allowing us to measure the influences of naturalness and usefulness upon image quality. To this end, we selected two kinds of manipulation; that is, varying the color temperature of the reference white[4] and varying chroma[5] (Hunt 1992), which we expected to affect naturalness and usefulness in distinctly different ways. The first manipulation was expected to influence only the naturalness of the image, whereas the second manipulation was expected to influence both the naturalness and the usefulness of the image.

Description

The experiment, similar to experiments described by Fedorovskaya et al. (1997), de Ridder, Fedorovskaya & Blommaert (1993), and de Ridder (1996), was performed using four color images of natural scenes taken from a Kodak Photo CD. The color temperature of the reference white was varied between 4,650 K and 10,300 K—6,500 K being the original—in seven steps of perceptually equal size, and chroma was scaled by a constant ranging from 0.5 to 2.0 in seven steps.

---

[4]Defined as "the temperature of a Planckian radiator whose radiation has the same chromaticity as that of a given stimulus." Unit: Kelvin (K).

[5]Defined as "the colorfulness of an area judged as a proportion of the brightness of a similarly illuminated area that appears to be white or highly transmitting." In the experiments C_{uv}^*, a correlate of chroma in the CIELUV color space, was scaled.

Seven subjects participated in the experiment. In three separate sessions they were shown on a CRT (cathode ray tube) the complete set of images, in random order, with three replications. In the first session, the subjects' task was to judge the quality[6] of the images, in the second session to judge the colorfulness[7] of the images, and in the third session to judge the naturalness[8] of the images. Subjects were instructed to use an 11-point numerical scale ranging from 0 ("bad" or "weak") to 10 ("excellent" or "strong").

Results

Colorfulness judgments (averaged over subjects and scenes) versus chroma (diamonds) and color temperature of the reference white (triangles) are shown in Fig. 3.2. The effect of scaling chroma on judged colorfulness is clearly visible. Less clear, although still significant, is the influence of color temperature of the reference white on judged colorfulness; images are judged less colorful for higher temperatures of the reference white.

Quality judgments (diamonds) and naturalness judgments (triangles) versus colorfulness judgments, all averaged over subjects and scenes, are shown in Fig. 3.3 for the conditions chroma (solid curves) and color temperature of the reference white (dashed curves). The figure shows that, as expected, quality correlates well with naturalness, although for the condition chroma the curve for quality is shifted with respect to the curve for naturalness toward higher values of colorfulness (and hence toward higher values of chroma). Note that a similar shift does *not* occur for the condition color temperature of the reference white. This seems to dismiss the possibility of a straightforward "preference" for images with higher colorfulness.

Quality judgments versus naturalness judgments, depicted in Fig. 3.4, again show that the quality–naturalness curve for the condition chroma deviates significantly from a linear relation. The curve has a U-formed shape resulting from the above-mentioned shift between quality judgments and naturalness judgments.

---

[6]Defined as "the degree to which you like the colors in the image."

[7]Defined as "the presence and vividness of the colors in the image."

[8]Defined as "the degree to which the colors in the image seem realistic to you."

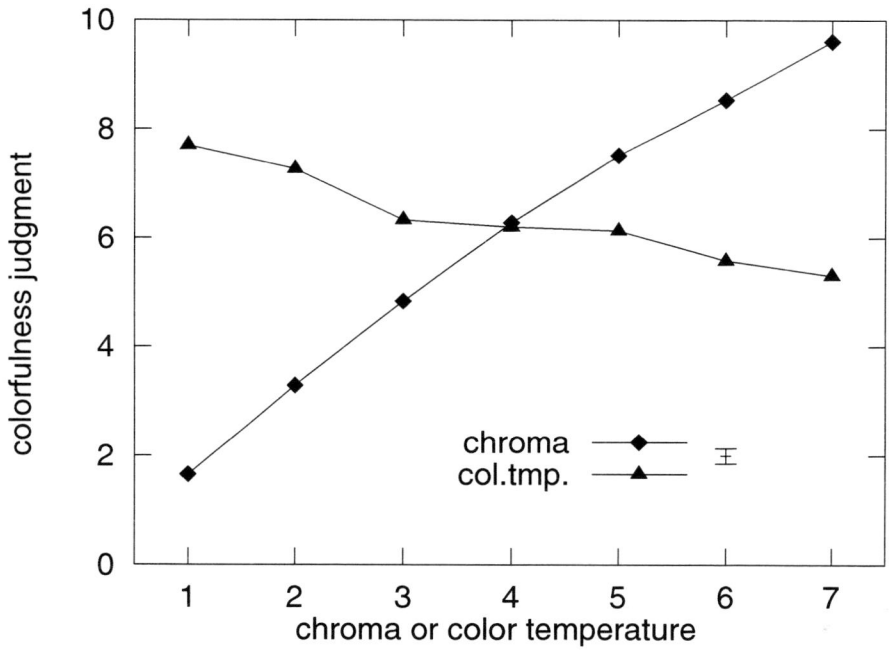

Figure 3.2: Colorfulness judgments (averaged over subjects and scenes) for the conditions chroma (diamonds) and color temperature of the reference white (triangles). The error-bar denotes a distance of two average standard errors in the mean. The numbers 1–7 on the horizontal axis denote, for chroma, scaling by 0.50, 0.63, 0.79, 1.00, 1.26, 1.59 and 2.00, and for the reference white a color temperature of 4,650 K; 5,150 K; 5,800 K; 6,500 K; 7,400 K; 8,650 K; and 10,300 K.

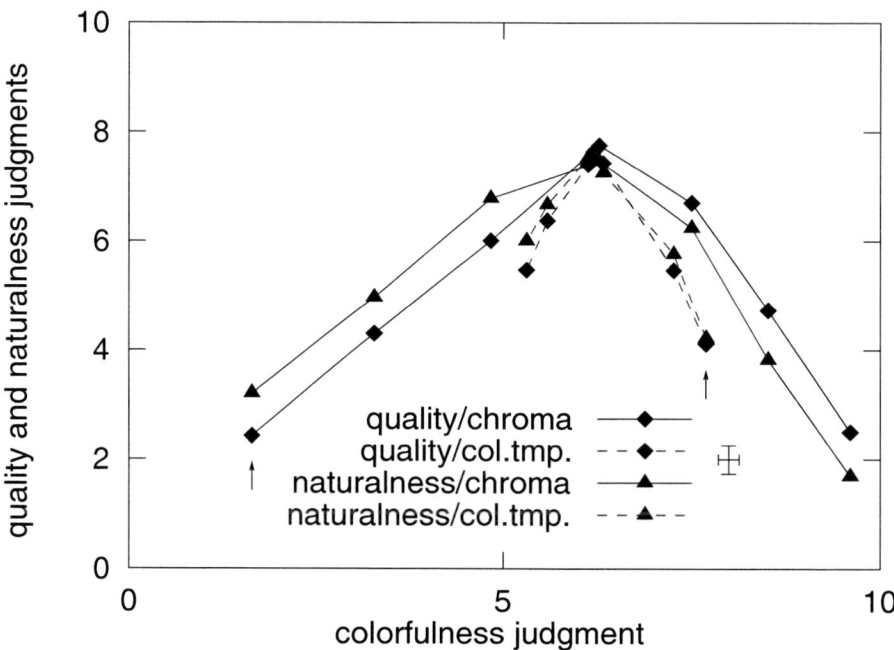

Figure 3.3: Quality judgments (diamonds) and naturalness judgments (triangles) versus colorfulness judgments (all averaged over subjects and scenes) for the conditions chroma (solid lines) and color temperature of the reference white (dashed curves). The error-cross denotes a distance of two average standard errors in the mean, and the arrows denote the images with lowest chroma and lowest color temperature of the reference white, respectively.

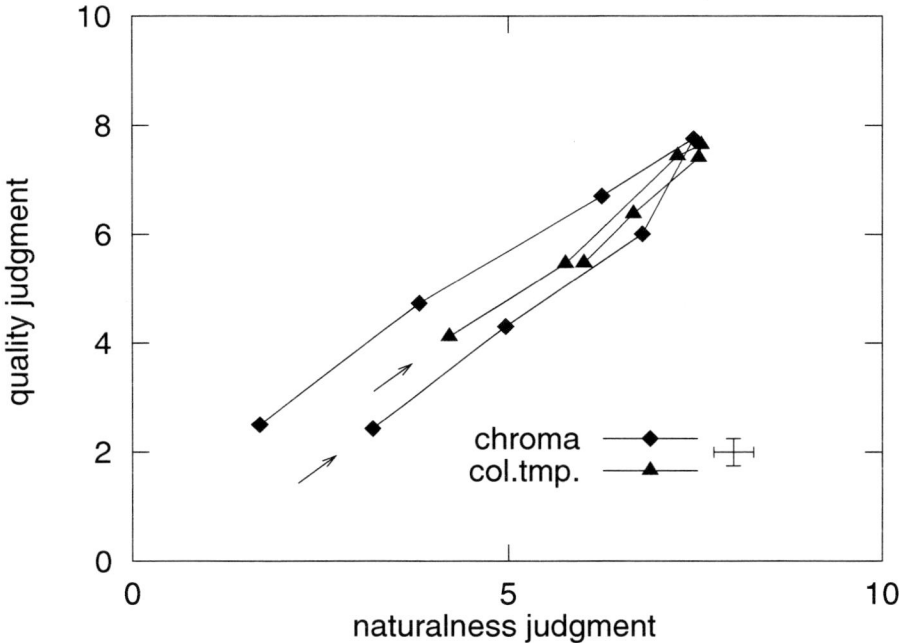

Figure 3.4: Quality judgments versus naturalness judgments (both averaged over subjects and scenes) for the conditions chroma (diamonds) and color temperature of the reference white (triangles). The error-cross denotes a distance of two average standard errors in the mean, and the arrows denote the images with lowest chroma and lowest color temperature of the reference white, respectively.

Interpretation

The first-order effect—that is, the high correlation ($r = 0.93$) between naturalness judgments and quality judgments for both conditions—can readily be interpreted in terms of the outlined semantic theory of image quality. In order to interpret the second-order effect—that is, the shift between naturalness judgments and quality judgments for the condition chroma—we adopt the CIELUV color space (recommended in 1979 by the Commission Internationale de l' Éclairage) as an appropriate, perceptually uniform color space.

In CIELUV, the image can be thought of as a cloud of dots, with each dot corresponding to one pixel in the image. Scaling chroma can be described as a radial contraction or expansion of the cloud of dots toward or away from the reference white, while changing the color temperature of the reference white can be described by a displacement of the entire cloud along the yellow–blue direction. The relevant difference between the two manipulations follows quite directly from their descriptions in CIELUV: changing chroma results in increased or decreased distances in color space between any pair of dots of which the members do not represent exactly the same color, while changing the color temperature of the reference white has no effect on these distances.

Contrary to manipulations that preserve distances in color space, manipulations that do affect distances in color space will also affect the precision with which the image can be represented internally (since at a presumed, constant level of internal noise, affecting distances in a perceptually uniform space is equivalent to affecting an internal "signal-to-noise ratio"), and hence the usefulness of the image. This discussion can therefore be concluded as follows: manipulations that do not affect the usefulness of an image—for example, changing the color temperature of the reference white—will have approximately identical parameter settings for optimizing the naturalness and the quality of the image. Manipulations that do affect the usefulness of an image—for example, changing chroma—will have different parameter settings for optimizing the naturalness or the quality of an image. In the latter situation, the parameter settings optimizing quality will tend to deviate with respect to those optimizing naturalness toward values that increase the usefulness of the image.

3.4.2 Experiment 2: Image quality regarded as a compromise between naturalness and usefulness

Aim

In order to test the description of image quality in terms of a compromise between naturalness and usefulness, we devised an experiment in which we manipulated the brightness contrast of black-and-white images of natural scenes. Assuming that (1) usefulness is linearly related to perceived brightness contrast; and (2) the compromise can be adequately described by a linear combination of naturalness and usefulness, we may write image quality Q in terms of naturalness N and brightness contrast C as

$$Q = \lambda_1 N + \lambda_2 C + \lambda_3, \tag{3.1}$$

and fit the vector $\vec{\lambda}$ to subjects' judgments of quality, naturalness and contrast as obtained in the experiment.

Description

The experiment was performed using four black-and-white images of natural scenes, obtained by transforming images from a Kodak Photo CD to the CIELUV color space and setting u^* and v^* to zero. We then applied the global, pixelwise transformation:

$$L^{*\prime} = \left(\frac{L^* - L^*_{min}}{L^*_{ave} - L^*_{min}} \right)^{\gamma} (L^*_{ave} - L^*_{min}) + L^*_{min}$$

$$(\text{for } L^*_{min} \leq L^* \leq L^*_{ave})$$

$$L^{*\prime} = \left(\frac{L^*_{max} - L^*}{L^*_{max} - L^*_{ave}} \right)^{\gamma} (L^*_{ave} - L^*_{max}) + L^*_{max}$$

$$(\text{for } L^*_{ave} < L^* \leq L^*_{max}), \tag{3.2}$$

(where L^* represents the original lightness value of a pixel, $L^{*'}$ its new value, and the subscripts "min," "max," and "ave" indicate the minimum, maximum, and average lightness values of the original image) on the image using for γ the values 0.25, 0.35, 0.50, 0.71, 1.00, 1.41, 2.00, 2.82, and 4.00. Applying this transformation will for $\gamma < 1$ decrease, and for $\gamma > 1$ increase the brightness contrast of the image. The minimum and maximum lightness of the image are not affected, while in general the average lightness will remain at approximately the same lightness value.

Eight subjects participated in the experiment. In three separate sessions they were shown on a CRT the complete set of images, in random order, with three replications. In the first session, the subjects' task was to judge the quality[9] of the image, in the second to judge the brightness contrast[10] of the images, and in the third session to judge the naturalness[11] of the images. Subjects were instructed to use an 11-point numerical scale ranging from 0 ("bad" or "low") to 10 ("excellent" or "high").

Results

Contrast judgments (averaged over subjects and scenes) versus the parameter γ are shown in Fig. 3.5. The figure shows that, as expected, contrast increases for increasing values of γ. Figure 3.6 shows quality judgments (diamonds, averaged over subjects and scenes) and naturalness judgments (triangles, also averaged) versus contrast judgments (also averaged). The curve for quality is shifted with respect to the curve for naturalness toward higher values of contrast; a result that is similar to the results for the condition chroma in experiment 1. To conclude, Fig. 3.7 shows quality judgments versus naturalness judgments. The U-formed shape already found in experiment 1 is clearly visible.

Interpretation

We again find a high ($r = 0.95$) correlation between naturalness judgments and quality judgments, confirming that naturalness is a principal factor

---

[9]Defined as "the degree to which you like the image."
[10]Defined as "apparent light-density differences."
[11]Defined as "the degree to which the image seems realistic to you."

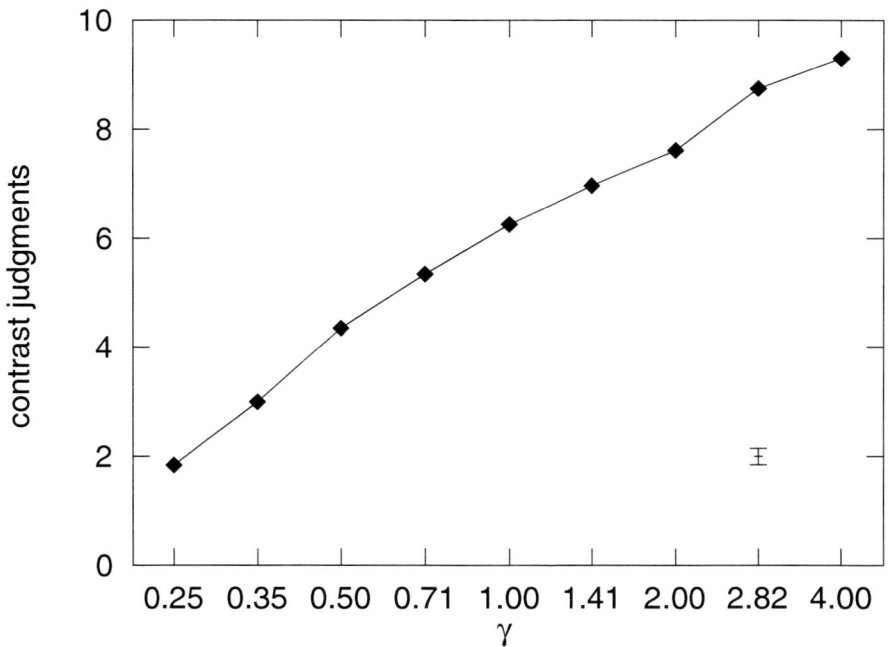

Figure 3.5: Contrast judgments (averaged over subjects and scenes) versus the parameter γ. The error-bar denotes a distance of two average standard errors in the mean.

constituting image quality. The shift of the curve for quality with respect to the curve for naturalness toward higher values of contrast can readily be interpreted when realizing that higher contrast allows for more accurate detection and localization of edges in the image, and thus for a more precise internal representation of the image.

The least-squares fit of our model to the quality judgments obtained in the experiment is given by

$$Q = \lambda_1 N + \lambda_2 C + \lambda_3$$
$$\vec{\lambda} = (0.90, 0.25, -1.08). \qquad (3.3)$$

Figure 3.8 shows this fit (circles), together with the quality (diamonds) and naturalness (triangles) judgments. The correlation between the fit and the

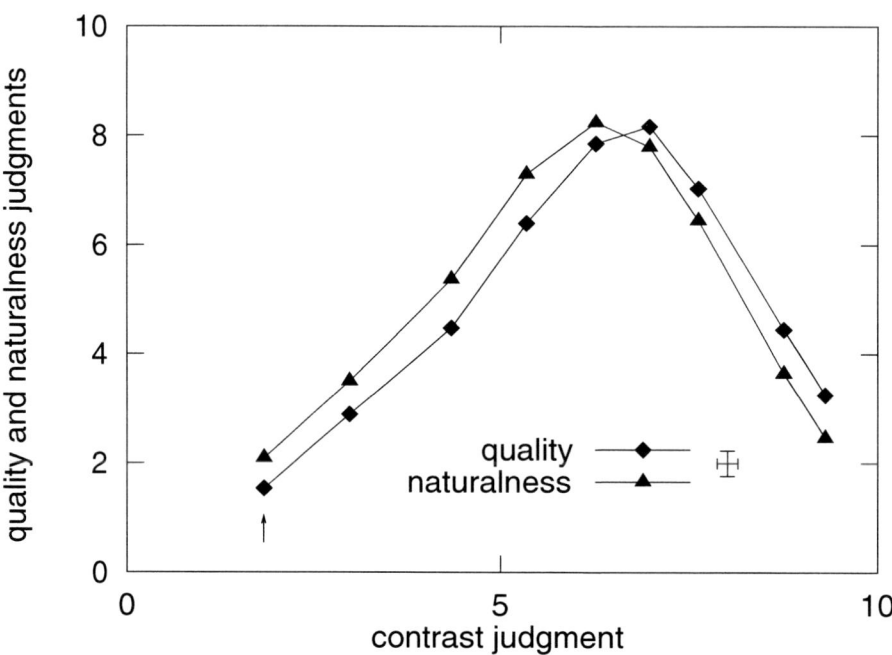

Figure 3.6: Quality judgments (diamonds) and naturalness judgments (triangles) versus contrast judgments (all averaged over subjects and scenes). The error cross denotes a distance of two average standard errors in the mean, and the arrow denotes the image with lowest γ.

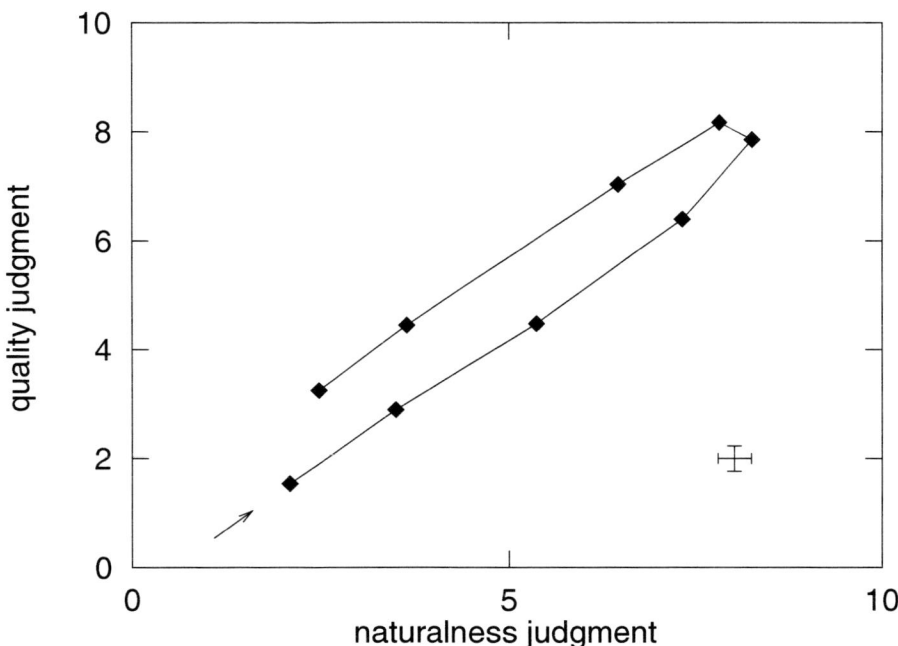

Figure 3.7: Quality judgments versus naturalness judgments (both averaged over subjects and scenes). The error-cross denotes a distance of two average standard errors in the mean, and the arrow denotes the image with lowest γ.

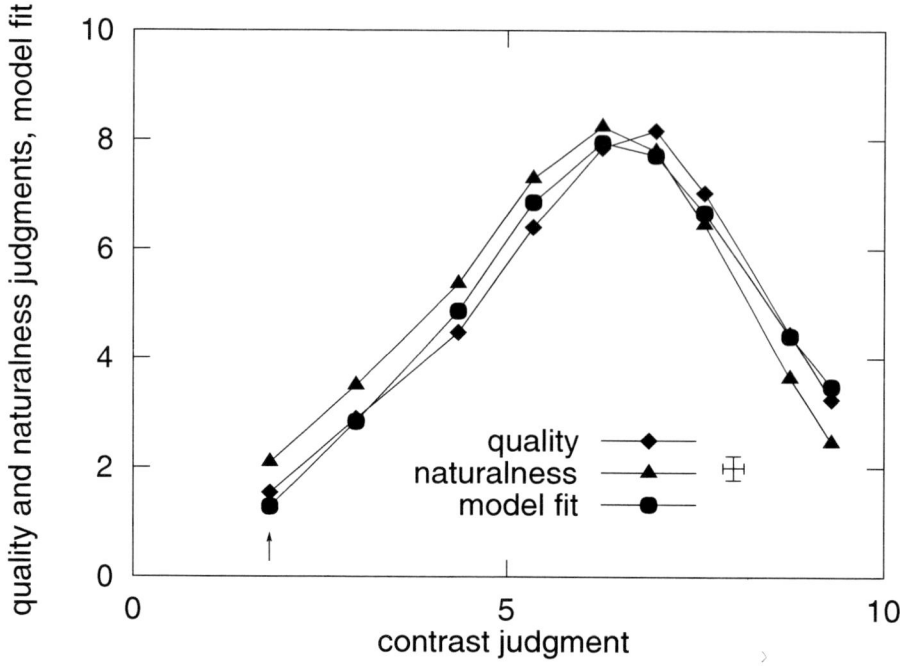

Figure 3.8: Quality judgments (diamonds) and naturalness judgments (tri-angles) versus contrast judgments (all averaged over subjects and scenes). The figure also shows the model fit (circles). The error cross denotes a distance of two average standard errors in the mean, and the arrow denotes the image with lowest γ.

quality judgments is very high ($r = 0.99$). Considering (1) the primitive nature of the model; and (2) the strong nonlinearity of the correlation co-efficient as a measure for goodness-of-fit (that is, goodness-of-fit increases strongly for correlation r approaching one), we may conclude that our description of image quality in terms of a compromise between naturalness and usefulness fits the data very well.

3.5 Concluding remarks

In pursuing a top-down, analytical approach, we have achieved a fundamental interpretation of the processes that play a role in the estimation of the attribute "quality" of an image. We have argued that image quality is, to the observer, a *useful* attribute of an image, expressing how well the observer is able to employ the image as a source of information about the outside world; a view of image quality that is strikingly different from the "perceived distance to the original" philosophy often employed in image quality research.

The results of the experiments discussed in the previous sections support the concept developed here that the quality of an image can be described in terms of a compromise between the naturalness and the usefulness of that image. A logical next step to proceed from this point onward would be to more thoroughly specify the naturalness and usefulness requirements imposed upon an image, for example, by means of formulating algorithms. Implementations of such algorithms will enable (1) the development of instrumental measures for the prediction of image quality; and (2) estimation of parameter settings optimizing the quality of images.

At this point, two generalizations of the ideas discussed here may be interesting to note. First, our description of quality is essentially formulated *independently* of modality, suggesting (1) the possibility of simply applying the same ideas to, for example, the fields of sound or speech quality; and (2) the possibility to generalize the current description of image quality to a multimodal semantic description of perceived quality of information presentation. It is highly likely that such a description will prove valuable in the design and evaluation of applications in which a multimodal presentation of information plays a central role.

Second, in our description of image quality we have concentrated on the requirements imposed upon the information that is acquired from the environment. However, requirements ensuring a proper interaction between observer and environment should necessarily include requirements that ensure the ability to adequately respond to the environment. These requirements may then, in general, be imposed upon the means the observer is employing to control his environment. Such an approach is likely to result in a general theory of the quality of man-machine interaction.[12]

---

[12]See Chapter 7 for a short discussion of two philosophical issues involved in this.

Chapter 4

Visual Metrics: Discriminative Power through Flexibility

An important stage in visual processing is the quantification of optical attributes of the outside world. We argue that the metrics used for this quantification are flexible, and that this flexibility is exploited to optimize the discriminative power of the metrics. We derive mathematical expressions for such optimal metrics and show that they exhibit properties resembling well-known visual phenomena. To conclude, we discuss some of the implications of flexible metrics for visual identification.

4.1 Introduction

Vision is often referred to as "inverse optics" (Poggio & Koch 1985), that is, the process of measuring the characteristics of an optical image of the environment and reconstructing the material properties of this environment. Defined this way, vision involves a stage in which optical attributes of the outside world and the objects located within it are measured and internally quantified. Examples of such internally quantified measures are position, shape, size, texture, color, and brightness. The metrics used for this quantification, in particular the metrics used for the quantification of color and brightness, will be the main topic of our discussion.

This chapter is a slightly modified version of Janssen & Blommaert (2000c), Visual metrics: discriminative power through flexibility, *Perception*, vol. 29, no. 8, pp. 965–980. Reprinted with permission of Pion Limited, London, UK.

First, to clarify what we mean here by the term metric, we will consider the measurement process.[1] To enable the measurement of the strength of a certain attribute, an *origin* and a *unit* must first be selected. For example, to measure temperature one might select the melting point of water for the origin, and select one hundredth of the temperature difference between the boiling point and the melting point of water for the unit. Unit and origin together define a *scale*, in this case the Celsius temperature scale. Thus, to measure the temperature of an item one would take the temperature difference between the item and the origin of the scale and express this difference in the number of temperature units. The number obtained in this way is the Celsius *scale value* for this particular temperature.

In the above example, equal differences in the attribute strength will lead to equal differences in the scale value. Scales having this property are referred to as linear scales, since the relation between attribute strength and scale value is linear. Most familiar scales are of this type; however, some well-known scales are not. Take, for example, the *dB* scale for sound pressure or the *pH* scale for the degree of acidity. These scales are known as logarithmic scales, since their unit is not defined in terms of a difference but instead in terms of a ratio. For logarithmic scales, equal ratios in the attribute strength lead to equal differences in the scale value, and the relation between attribute strength and scale value is therefore logarithmic. For a certain scale, the exact relation between attribute strength and scale value is made explicit by the *scale function*. The scale function, together with the scale, the origin, and the unit, constitutes what is called a *metric* (Watt 1989).

Standardized metrics as the Celsius, *dB*, or *pH* metrics share one important and useful property: rigidity. Rigidity refers to the property that the scale function is uniquely defined and constant throughout time. For most measurements this property is essential. The Celsius temperature scale would be of little value when a temperature of 37°C as measured today would be different from a temperature of 37°C as measured tomorrow. At least, predicting tomorrow's temperature in terms of the Celsius temperature scale

---

[1]Throughout this chapter we will use the following definition of the term metric: *a metric is the instrument for the quantification of measurement results.* Another widely known definition of the term metric, one which we will *not* be using, is given in terms of a non-negative, real-valued distance function $d(.)$ [this distance function must satisfy the three requirements $d(x, x) = 0$, $d(x, y) = d(y, x)$, and $d(x, z) \leq d(x, y) + d(y, z)$].

would make little sense. Rigidity of a metric allows for a unique specification of the attribute strength in terms of the scale value; 37°C is one and the same temperature, whenever and wherever you measure it.

We now return to vision and visual metrics.[2] Traditionally, the view in visual research has been that visual metrics, such as the brightness metric, are essentially rigid (Weber 1846, Fechner 1860, Riesz 1933, Stevens 1957). Several exceptions to this rigidity are well known, for instance dark and light adaptation (Cohn & Lasley 1986) and crispening (Takasaki 1966) for the case of brightness; however, each of these phenomena has traditionally been described and modeled separately. As far as we know, there have been few attempts to unify these phenomena into one consistent description.

What we will try to do here is to follow an approach in which we will regard visual metrics no longer as being rigid.[3] More specifically, we will assume that visual metrics are (1) limited in range; that is, the scale has fixed lower and upper bounds; (2) limited in accuracy; that is, scale values cannot be represented with arbitrary precision due to the presence of noise; (3) intrinsically *flexible* (Blommaert 1995), which means that the scale function is allowed to vary in time; and (4) optimized with respect to overall *discriminative power* (Blommaert 1995, Watt 1989, Watt 1991), which means that the scale function is chosen such that the ability to discriminate between items in the outside world using the scale values of their measured attributes, is maximized.

We will divide our discussion of flexible metrics into four parts. First, we will consider the above four assumptions and discuss how and under

---

[2]Visual metrics, and sensory metrics in general, are often referred to as *sensory scales*. The topic of sensory scales has traditionally generated a large amount of literature; for some recent work see Falmagne (1985), Laming (1986), Luce & Narens (1987), and Gescheider (1988). Some monumental contributions are by Weber (1846), Fechner (1860), and Stevens (1957). Assuming that the scale function is rigid, much attention has been given to the exact shape of the scale function; in particular, whether this shape can be described by a logarithm (Weber 1846, Fechner 1860) or by a power-law (Stevens 1957). In a nutshell, the question we try to address in this manuscript is the following: given that visual metrics serve the purpose of quantifying outside world attributes, and given that the aim of this quantification is to discriminate items in the outside world, what would be the *optimal* shape of the scale function? As will be shown, this optimal shape requires the scale function to be *flexible*.

[3]A classical example of this approach, applied to the metrics of visual space, is given by Andrews (1964). A similar approach is also followed by Watt (1989).

41

what circumstances flexibility can be exploited to improve the discriminative power of a metric. Second, we will derive mathematical expressions for metrics satisfying the above assumptions. Third, using the obtained expressions we will explore some of the properties of these metrics, and relate these properties to well-known visual phenomena, namely brightness constancy (Wallach 1948) and crispening (Takasaki 1966). Last, we will consider some of the consequences of flexibility for the process of visual identification.

4.2   The usefulness of flexibility

We have assumed that visual metrics are limited in range and accuracy and, most importantly, that they are flexible and optimal with respect to discriminative power. The assumptions of limited range and limited accuracy seem straightforward, certainly when the constraints imposed by physical or biological implementations of these metrics are taken into account. Flexibility, however, is a less straightforward assumption. Flexibility may or may not prove useful, depending on the type of measurement that is performed. Noticeably, a measure requiring a strict one-to-one correspondence of scale value to attribute strength, as in the example of the temperature scale, will leave little space for flexibility of the metric.

The goal of early visual processing, however, is to extract the maximum possible amount of the information contained within the optical image, and to represent this information in a way that maximizes its usefulness to the organism. Here, usefulness does not necessarily imply a one-to-one correspondence with reality. Instead, a more meaningful criterion for usefulness is often the *ability to discriminate* between items in the outside world based on information about these items as it is represented internally. For example, the ability to discriminate between items on the basis of their colors may often be more important than establishing the exact colors of these items. For the aim of discriminative power, a one-to-one correspondence between internal representation and outside world is unnecessary. In fact, it is likely that such strict correspondence only decreases the ability to discriminate.

How can flexibility be exploited to improve discriminative power? To answer this question we will consider the assumptions of limited range and accuracy again. When scale values cannot be represented with arbitrary precision, for example, due to the presence of noise, the ability to discriminate items in the outside world using their scale values will be essentially limited by the noise level. Items will be discriminable only when their scale values are at least on the order of one noise level apart, and, assuming the noise level to be constant, discriminability will increase monotonically with scale value difference. Ultimately, to increase the overall discriminative power of a metric all scale value differences should be increased. Again assuming the noise level to be constant, this could be done by "stretching" the entire scale to a larger range. However, this range was assumed to be limited and constant.

Alternatively, overall discriminative power may be increased by *locally* stretching the scale, by means of locally increasing the derivative of the scale function, and by compressing the scale elsewhere. Such a mechanism allows for increased scale value differences while it simultaneously preserves the range of the metric. When the locations where the scale is stretched or compressed are carefully chosen, this may lead to a significant increase in overall discriminative power. This principle is illustrated in Fig. 4.1. The figure shows a simplified situation for the example of brightness (the internal measure) versus luminance (the attribute strength, given on a logarithmic scale). The situations for rigid and flexible metrics are shown in the left and right panel, respectively. The open circles on the luminance axes denote the luminances of a set of items under daylight illumination. The filled circles denote the luminances of the same set of items, this time under nocturnal illumination conditions.

First, consider the left panel of Fig. 4.1. For a rigid metric to be able to represent all possible attribute strengths, the entire range of the attribute strength somehow has to be mapped onto the internal scale. For the case of luminance, the range of the attribute strength is estimated to be 10 orders of magnitude; however, at any particular moment the luminance range found is typically only a small fraction of this entire possible range, usually only about three orders of magnitude (McCann 1988). Thus, even if the scale function of our rigid metric was logarithmic, only about 30 percent of the metric would really be used at any particular moment. Boldly stated, this would be a waste of resources.

By locally stretching the scale we might increase all scale value differences by a factor of up to three, thereby significantly increasing overall discrim-

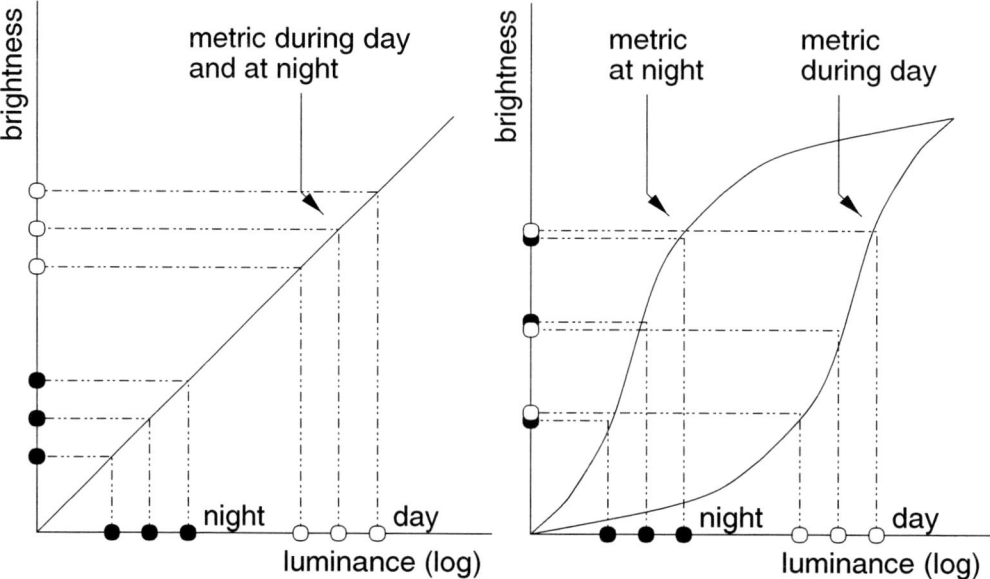

Figure 4.1: Rigid metric versus flexible metric. Simplified case for brightness (internal measure) versus luminance (physical attribute, on a logarithmic scale). Shown are the luminances and corresponding brightnesses of a set of items under daylight (open circles) and nocturnal (filled circles) illumination conditions.

inative power at any particular moment. However, to do this effectively, the locations where to stretch the scale need to depend on the current range of the attribute strength. Intuitively, the conclusion may already be that the amount of stretching of the scale must be closely related to the momentary distribution of the attribute strength, as shown in the right panel of Fig. 4.1. Here we encounter flexibility again, since in this case the scale function must be allowed to "follow" fluctuations in the momentary distribution of the attribute strength.

In the above discussion we have assumed the presence of noise, and it is therefore important to characterize the sources of this noise in more detail. First, we distinguish noise sources acting directly upon the attribute strength. For the case of luminance such a noise source would be photon noise. Second, we distinguish noise sources acting directly upon the scale value. When scale values are assumed to be encoded in neuronal impulse rates, random variability in these rates may be regarded as such a noise source. The first type of noise typically originates outside of the "measuring device," whereas the second type is generated by the measuring device itself. We will therefore refer to the first and second type of noise as *external* and *internal* noise, respectively. Figure 4.2 gives a graphical summary of this. Both internal and external types of noise can be regarded as compound noise; that is, each type can be thought to consist of a number of independent contributions from different sources. Usually, compound noise can be adequately modeled by assuming that the noise has Gaussian properties. This assumption will also facilitate the mathematical description in the next section, since Gaussian noise can be completely specified in terms of the two parameters mean and standard deviation.

Before we can start to derive mathematical expressions for optimal metrics, we will need to define a measure for overall discriminative power. The measure we will use here is the total number of *topological errors* made in the mapping of attribute strength onto scale value. Such topological errors occur when the ordering of a set of items by their scale values differs from the ordering of this same set of items by their attribute strength. For example, if $x_1 < x_2 < x_3 < x_4 < x_5$ were to be the result of ordering a set of five items by their attribute strengths x_i (where $i = 1, \ldots, 5$), and if $s_1 < s_4 < s_3 < s_2 < s_5$ were to be the result of ordering the same set of items by their scale values s_i, the resulting number of topological errors would be three ($s_4 < s_3$, $s_3 < s_2$, and $s_4 < s_2$; whereas $x_3 < x_4$, $x_2 < x_3$, and $x_2 < x_4$). The main

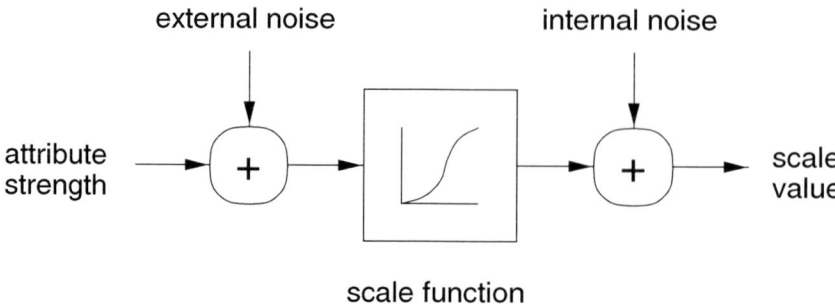

Figure 4.2: Block diagram showing the definitions we use here of external and internal noise.

justifications for choosing the number of topological errors to be our discriminability measure are (1) the number of topological errors is likely to be highly correlated with discriminative power, since when topological errors occur, discriminative power *must* be poor; and (2) the number of topological errors is an intuitive and clear measure to calculate. To conclude, note that the exact shape of the discriminability measure is not very important, provided that (1) discriminability is poor for scale value differences on the order of one noise level or less; and (2) the noise level is small compared to the range of the scale.

4.3   Recipe for an optimal metric

4.3.1   Problem specification

Assume that, at some moment in time, the values of the attribute strength x (for example, luminance) are measured for a given set of N items, where the set of N items is chosen as an abstraction for the environment. If the set of items is large, we may approximate the momentary distribution of the measured attribute strengths for the set by a continuous function $\mu(x)$, such that the number of measured attribute strengths dN within an arbitrary small interval $[x, x + dx]$ is given by $dN = \mu(x)dx$. Note that, according to this definition, $\int_X \mu(x)dx = N$, where X is the range of the attribute strength.

For the example of luminance, the N items can be thought of as the pixels constituting an image, while $\mu(x)$ can be thought of as the luminance frequency distribution of the image, scaled with a factor N.

Furthermore, assume that the measurement results are represented by scale values on an internal scale s that is monotonic with x; that is, $s = s(x)$ and $ds(x)/dx \geq 0$ for all x, where $s(x)$ is the scale function. Notice that monotonicity is required here, since any deviations from monotonicity will inevitably result in topological errors being made in the mapping from x to s. The momentary distribution of the scale values on s for a given distribution $\mu(x)$ of the attribute strengths and for a given scale function $s(x)$ will be referred to as $\eta(s)$. If we want to optimize the scale s such that overall discriminability of the items on s is maximized, an obvious strategy here is to find the scale function $s(x)$ that minimizes the number of topological errors.

The solution to this problem is as follows. Consider a small interval $[s, s + ds]$ on the scale s; the number of scale values in the interval $[s, s + ds]$ is then given by $dN = \eta(s)ds$. When the probability of a topological error is written as a function of s, that is, $p_e = p_e(s)$, we may write the number of topological errors in the interval $[s, s + ds]$ as

$$dN_e = p_e(s)\eta(s)ds. \tag{4.1}$$

The total number of topological errors, N_e, is found by integrating this expression along the entire scale. Assuming that the range of the scale is given by S, we thus find

$$N_e = \int_S p_e(s)\eta(s)ds. \tag{4.2}$$

Based on the assumption that the noise limiting the precision with which scale values can be represented is Gaussian, we show in Appendix 4.A that the probability of a topological error can be expressed as a monotonic function[4] of the total noise level $\sigma(s)$ times the item density $\eta(s)$:

$$p_e(s) = \frac{1}{2} - \frac{1}{2} \operatorname{erf} \frac{1}{2\eta(s)\sigma(s)}. \tag{4.3}$$

---

[4]Figure 5.4 in Chapter 5 shows a plot of $p_e(s)$ versus $d/\sigma(s)$ [where $d = 1/\eta(s)$].

Assuming Gaussian noise propagation, the total noise level in Eq. (4.3) can be expressed in terms of the internal noise level $\sigma_i(s)$ and the external noise level $\sigma_e[x(s)]$ as:

$$\sigma(s)^2 = \sigma_i(s)^2 + \sigma_e[x(s)]^2 \left[\frac{ds(x)}{dx}\right]^2. \tag{4.4}$$

Finally, the link between $\eta(s)$ and $\mu(x)$ has to be made to solve the problem. This link follows from $s(x)$ being monotonic and the number of items being preserved from an interval $[x, x+dx]$ to the corresponding interval $[s, s+ds]$, that is:

$$\eta(s)ds = \mu(x)dx. \tag{4.5}$$

The optimal scale s is found by substituting Eqs. (4.3), (4.4), and (4.5) into Eq. (4.2) and minimizing the resulting equation for $ds(x)/dx$.

Before trying to solve the set of Eqs. (4.2)–(4.5), we will first consider an important consequence of Eq. (4.4). We have argued that locally stretching the scale can be used to increase discriminability. By regarding Eq. (4.4) we may investigate under what circumstances this mechanism will work. To this end, consider that the derivative of the scale function, $ds(x)/dx$, is used here as the instrument to increase the scale value difference ds of two items with attribute strength difference dx:

$$ds = \left[\frac{ds(x)}{dx}\right] dx. \tag{4.6}$$

Discriminability of these two items will be determined by the *ratio* of scale value difference to the total noise level (SNR):

$$\text{SNR} = \frac{ds}{\sigma(s)}$$

$$= \frac{1}{\sigma(s)} \left[\frac{ds(x)}{dx}\right] dx, \tag{4.7}$$

where $\sigma(s)$ is given by Eq. (4.4). We now distinguish between two extreme cases: (1) external noise dominant; and (2) internal noise dominant. For the first case, Eq. (4.4) can be approximated by

$$\sigma(s) = \sigma_e[x(s)] \left[\frac{ds(x)}{dx} \right]. \tag{4.8}$$

The ratio of scale value difference to total noise level then reduces to

$$\text{SNR} = \frac{dx}{\sigma_e[x(s)]}, \tag{4.9}$$

which is *independent* of $ds(x)/dx$. Therefore, when the influence of external noise is dominant, discriminability *cannot* be improved by stretching of the scale.

For the second case, Eq. (4.4) can be approximated by

$$\sigma(s) = \sigma_i(s), \tag{4.10}$$

and Eq. (4.7) reduces to

$$\text{SNR} = \left[\frac{ds(x)}{dx} \right] \frac{dx}{\sigma_i(s)}. \tag{4.11}$$

Here, SNR increases proportionally with the amount of stretching of the scale, and therefore stretching of the scale can indeed be used as a mechanism to improve discriminability. Having observed this, we will in the remainder of our discussion assume that the influence of internal noise is dominant; that is, we will assume that Eq. (4.4) can be approximated by Eq. (4.10). In some situations this assumption may not be true, for example, at very low illumination levels where photon noise becomes important.

4.3.2 Solution

We start deriving the solution to our problem by regarding the Maclaurin series of the error function:

$$\text{erf } z = \frac{2}{\sqrt{\pi}} \sum_{m=0}^{\infty} \frac{z^{2m+1}}{m!(2m+1)}. \tag{4.12}$$

Substituting $z = 1/[2\eta(s)\sigma(s)]$ in Eq. (4.12) and substituting the result in Eq. (4.3), we get:

$$p_e(s) = \frac{1}{2} - \frac{1}{\sqrt{\pi}} \sum_{m=0}^{\infty} \frac{[2\eta(s)\sigma(s)]^{-(2m+1)}}{m!(2m+1)}. \tag{4.13}$$

Substituting the obtained expression for $p_e(s)$ in Eq. (4.2), we find for the total number of topological errors N_e:

$$
\begin{aligned}
N_e &= \int_S \left\{ \frac{1}{2} - \frac{1}{\sqrt{\pi}} \sum_{m=0}^{\infty} \frac{[2\eta(s)\sigma(s)]^{-(2m+1)}}{m!(2m+1)} \right\} \eta(s)ds \\
&= \frac{1}{2} \int_S \eta(s)ds - \frac{1}{\sqrt{\pi}} \sum_{m=0}^{\infty} \int_S \frac{[2\eta(s)\sigma(s)]^{-(2m+1)}\eta(s)}{m!(2m+1)} ds. \tag{4.14}
\end{aligned}
$$

The first term of Eq. (4.14) is constant, since $\eta(s)$ integrated over the range S should always be equal to N, the number of items. As can be concluded from Eq. (4.14), the contribution dN_e of each individual term in the sum of Eq. (4.14) to the total number of errors N_e has the general form

$$dN_e = c_n \int_S \eta(s)^{n+1} \sigma(s)^n ds, \tag{4.15}$$

where $n = -(2m+1)$ and all c_n are negative. We will first concentrate on finding a solution for this general form, and then infer the solution of Eq. (4.14) from it. When Eq. (4.5) is substituted in Eq. (4.15), the latter can be written as

$$dN_e = c_n \int_X \sigma[s(x)]^n \mu(x)^{n+1} \left[\frac{ds(x)}{dx}\right]^{-n} dx, \tag{4.16}$$

where X is the range of the attribute strength. Assuming that internal and external noise are additive, the noise levels will be constant; that is, $\sigma_i(s) = \sigma_i$ and $\sigma_e(x) = \sigma_e$. Using Eq. (4.4), we obtain:

$$dN_e = c_n \int_X \left\{ \sigma_i^2 + \left[\frac{ds(x)}{dx}\right]^2 \sigma_e^2 \right\}^{n/2} \mu(x)^{n+1} \left[\frac{ds(x)}{dx}\right]^{-n} dx, \tag{4.17}$$

which should be minimized for $ds(x)/dx$. As discussed at the end of the previous section, we will look at the solution of Eq. (4.17) for the case that internal noise is dominant; that is, we assume that $\sigma_i \gg [ds(x)/dx]\sigma_e$ and approximate Eq. (4.4) by Eq. (4.10). For Eq. (4.17) we then find

$$
\begin{aligned}
dN_e &= c_n \int_X \sigma_i^n \mu(x)^{n+1} \left[\frac{ds(x)}{dx} \right]^{-n} dx \\
&= c_n \sigma_i^n \int_X \mu(x)^{n+1} \left[\frac{ds(x)}{dx} \right]^{-n} dx,
\end{aligned}
\tag{4.18}
$$

since σ_i was assumed to be constant. The problem of minimizing Eq. (4.18) is an example of the general problem of finding the extremes of the integral

$$
J = \int_{x_1}^{x_2} \phi(x, y, y') dx,
\tag{4.19}
$$

where $y = y(x)$ is some function of x, $y' = dy(x)/dx$, and ϕ is a function of the variables x, y, and y'. The integral of Eq. (4.19) can be interpreted as a line integral along some path $\Gamma = y(x)$ joining the points (x_1, y_1) and (x_2, y_2), see, for example, Irving & Mullineux (1959, pp. 362 ff.). Therefore, the problem is to find the path $\Gamma_0 = y_0(x)$ that extremizes the integral J. The solution to this problem is given by the *Euler-Lagrange equation*:

$$
\frac{\partial \phi}{\partial y} - \frac{d}{dx} \left(\frac{\partial \phi}{\partial y'} \right) = 0,
\tag{4.20}
$$

which, when ϕ does not contain y explicitly, reduces to

$$
\frac{d}{dx} \left(\frac{\partial \phi}{\partial y'} \right) = 0.
\tag{4.21}
$$

Integrating Eq. (4.21) with respect to x yields

$$
\frac{\partial \phi}{\partial y'} = k,
\tag{4.22}
$$

where k is a constant. After substitution of $y' = ds(x)/dx$ and of $\phi = \mu(x)^{n+1}[ds(x)/dx]^{-n}$ into Eq. (4.22), we obtain the solution to our problem:

$$\frac{ds(x)}{dx} = k'\mu(x). \tag{4.23}$$

By integrating Eq. (4.23) over the range X, it can easily be shown that the constant k' must be equal to S/N. Interestingly, the solution we have found is independent of the value of n, which means that we have found *one* solution that extremizes *every* individual term in the sum of Eq. (4.14). The entire sum will therefore be extremized as well, and Eq. (4.23) represents the solution to the overall problem of minimizing N_e in Eq. (4.14).

The solution we have found here resembles *histogram equalization* of $\mu(x)$ since, substituting Eq. (4.23) into Eq. (4.5):

$$
\begin{aligned}
\eta(s) &= \mu(x)\left[\frac{ds(x)}{dx}\right]^{-1} \\[2mm]
&= \frac{N}{S} \\[2mm]
&= \text{constant.} \tag{4.24}
\end{aligned}
$$

Histogram equalization is a well-known image processing tool used to increase image contrast or detectability of image features; see, for example, Ballard & Brown (1982, pp. 70–72). The principal difference between the solution we find here and standard histogram equalization is that here the characteristics of a flexible "measuring" device are adapted to the characteristics of the input signal, whereas in histogram equalization the characteristics of the input signal are adapted to a supposedly rigid measuring device.

Furthermore, it is a well-known result from information theory that the entropy of a range-limited stochastic signal is maximal when the probability density function of this signal is uniform; see, for example, Thomas (1969, pp. 561–563). It is interesting to note that an optimal scale function $s(x)$, as defined by us in terms of discriminability, maps an arbitrary distribution $\mu(x)$ to a uniform distribution $\eta(s)$, thereby maximizing the entropy of the distribution. In other words, the scale function that optimizes the discriminative power of a metric is the same as the mapping function needed in nonlinear recoding to optimize the amount of transmitted information. Barlow (1961) discusses such redundancy-reducing recoding of sensory information as one of the hypothetical functions of the sensory systems.

4.3.3 Flexibility versus rigidity: performances compared

How much can be gained by using flexible metrics? To obtain an impression of this, we used for the momentary distribution $\mu(x)$ the calculated luminance distributions of 40 digitized images of natural scenes taken from a Kodak Photo CD, assuming that these images were to be displayed on a CRT with $\gamma = 2.5$. We then assumed three types of brightness metric: (1) a linear metric; that is, a metric of the type $s(x) = x/x_0$, where x_0 is an arbitrary reference value such that $x/x_0 \leq 1$; (2) a compressive metric; that is, a metric of the type $s(x) = (1 + \beta)(x/x_0)^{1/3} - \beta$ [which for $\beta = 0.16$ is proportional to CIE 1976 lightness L^*; see Hunt (1992, p. 72)]; (3) an optimal metric; that is, a metric according to Eq. (4.23). The optimal metric therefore varied from image to image, whereas the other metrics remained invariant. To facilitate a comparison of the performances of the three metric types, their ranges were chosen such that $s(x) \in S = [0, 1]$ for all $x \in X = [0, x_0]$.

We then calculated, using Eqs. (4.2), (4.3), and (4.5), for each image and each type of metric the number of topological errors N_e. To this end, we assumed that the total noise level $\sigma(s)$ is dominated by a constant internal noise level σ_i equal to one percent of the range of the metric; that is, we assumed that $\sigma(s) = \sigma_i = 0.01S = 0.01$ (corresponding to a dynamic range of approximately 100 JNDs). Furthermore, we scaled the distributions $\eta(s)$ by a factor of 100, thereby assuming that each image contained exactly 100 items to be discriminated.

Figure 4.3 shows frequency distributions of the number of topological errors for the linear, compressive, and optimal metrics. Note that the number of topological errors for the optimal metric is the same for all images, something that is explained by $\eta(s)$ being constant and independent of $\mu(x)$ for optimal metrics. Figure 4.3 shows that, as expected, the optimal metric out performs the two rigid metrics; the average number of topological errors for the linear, compressive, and optimal metrics is 34.0 ± 3.8, 29.3 ± 2.2, and 24.0, respectively.

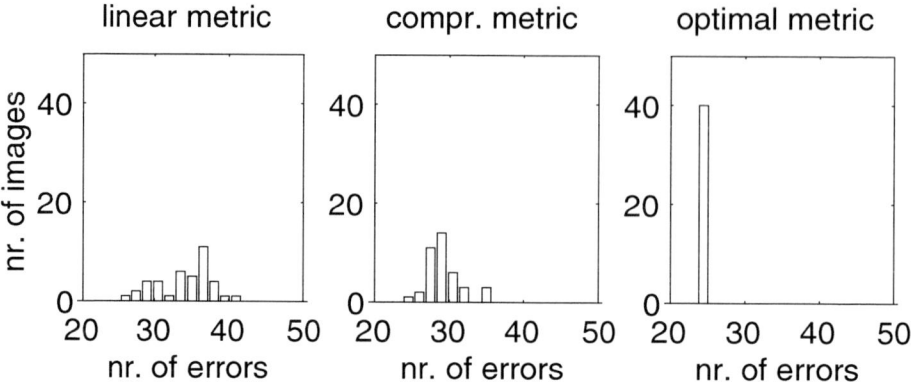

Figure 4.3: Discriminability, expressed in the number of topological errors, calculated for 40 images of natural scenes and for three types of metric: linear, compressive and optimal.

It is important to realize here that the above performance difference as calculated for digitized images is still relatively low, since the luminance ranges of the individual images are mapped into approximately the same range as a result of nearly optimal choices for diaphragm and exposure times when the images were taken. It is in everyday situations, where luminance ranges differ enormously, where the improved performance of flexible metrics becomes a distinct advantage.

4.3.4   Concluding remarks

Two important issues relating to the momentary distribution $\mu(x)$ of the attribute strength x have so far not been addressed. The first issue is related to the influence of the factors time and location on $\mu(x)$. Consider, for example, the influence of time. According to the ideas we have forwarded, the scale function of a flexible metric should instantaneously follow any changes in the momentary distribution of the attribute strength. However, adaptation in real vision is known to be much faster when the illumination level increases than when the illumination level decreases. Although this asymmetry may simply be due to constraints in the biological implementation of the adaptation process, it may also serve to reduce the risk of physical damage resulting from exposure to high illumination levels.

The second issue is related to the concept of $\mu(x)$ itself, which is ill defined in the context of real vision. We have introduced $\mu(x)$ as the momentary distribution of the attribute strength of "a set of items." The question arises what the correlate of $\mu(x)$ may be in real vision. Considering that flexibility can be most successfully exploited during the earliest stages of visual processing, where measures of attribute strength are least affected by the cumulative influence of internal noise, we may conclude that it would have to be a very early visual property.

Notwithstanding these issues, many well-known properties of vision fit in very well with the description of optimal visual metrics we have derived. For example, it can easily be shown that optimal metrics as specified by Eq. (4.23) exhibit properties that resemble phenomena such as dark and light adaptation (Cohn & Lasley 1986), brightness constancy (Wallach 1948), and crispening (Takasaki 1966). First, to demonstrate the case for brightness constancy, consider that when the momentary luminance distribution is given by $\mu(x)$, the brightness s_0 of an item with luminance x_0 is given by

$$
\begin{aligned}
s_0 &= \int_{-\infty}^{x_0} \frac{ds(x)}{dx} dx \\
&= \frac{S}{N} \int_{-\infty}^{x_0} \mu(x) dx.
\end{aligned} \tag{4.25}
$$

When the illumination level changes by a factor c, all luminances will increase by this factor, since luminance is the product of illuminance and surface reflectance, and surface reflectance remains constant. We therefore find that the new luminance x_0' of our item will be equal to cx_0, and, in analogy, that the new luminance distribution $\mu'(x)$ will be equal to $1/c \cdot \mu(x/c)$. The new brightness s_0' of our item is given by

$$
\begin{aligned}
s_0' &= \frac{S}{N} \int_{-\infty}^{x_0'} \mu'(x) dx \\
&= \frac{S}{N} \int_{-\infty}^{cx_0} \frac{1}{c} \mu\left(\frac{x}{c}\right) dx \\
&= \frac{S}{N} \int_{-\infty}^{x_0} \mu(x') dx' \\
&= s_0,
\end{aligned} \tag{4.26}
$$

that is, the brightness of our item is not influenced by the illumination level.

Second, to demonstrate the case for crispening, consider that for a small luminance difference dx the corresponding brightness difference ds is given by

$$ds = \left[\frac{ds(x)}{dx} \right] dx$$

$$= \frac{S}{N} \mu(x) dx. \qquad (4.27)$$

The background luminance x_0 will usually appear in the luminance distribution $\mu(x)$ as a distinct peak for $x = x_0$. We therefore find that for a constant luminance difference dx, the associated brightness difference ds will also peak around $x = x_0$. In other words, we find that sensitivity for luminance differences is highest around the background luminance.

To conclude this discussion, we would like to stress that the above phenomena can now be understood as the *logical consequences* of visual metrics having one essential property which itself is extremely useful: continuous optimization of discriminative power by exploiting flexibility.

4.4 Vision and visual memory

4.4.1 Visual identification: vision versus memory

The ability to discriminate between items based on their visual attributes is essential to any seeing organism. However, advanced vision is also characterized by the ability to identify or recognize items or item properties using past observations of these items. For this process we will assume the simple model shown in Fig. 4.4, in which identification is performed by comparing, or matching, scale values of measured item attributes with "standards" stored in memory.

The obvious way in which these standards are themselves constructed in memory is by the accumulation of past observations of the items. Such a long-term temporal integration will inevitably result in memory standards

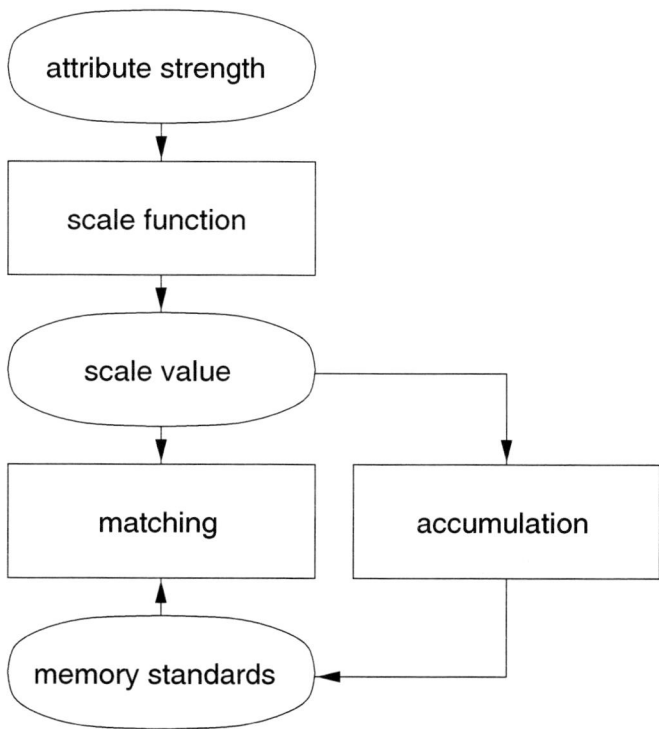

Figure 4.4: A simple model for the identification of items in the outside world. Identification is assumed to be performed by means of matching scale values with standards stored in memory. The memory standards themselves are assumed to be constructed by means of accumulation of scale values of past observations.

that are essentially rigid, at least when regarded in short-term intervals. This finding immediately raises an interesting question: how are observations, represented by scale values on a *flexible* visual metric, to be compared with standards that are essentially *rigid*?

4.4.2 Calibrating visual metrics

One way to cope with the above problem of comparing measurements represented on a flexible visual metric with rigid memory standards is to "calibrate" the visual metric. Such calibration can be performed in several ways. We distinguish four possible calibration methods:

- First, the visual metric can be calibrated when the noise properties, or estimates of these, are known. Referring to Eq. (4.4), the metric can be found using

$$\left[\frac{ds(x)}{dx}\right]^2 = \frac{\sigma(s)^2 - \sigma_i(s)^2}{\sigma_e[x(s)]^2}. \tag{4.28}$$

- Second, the attributes of already recognized items can be used to estimate the current visual metric. This method explicitly requires that other item attributes be used to recognize these items first. For example, shape might be used to identify several items first, after which their "known" colors could be used to calibrate the metric for color.

- Third, statistical properties of the momentary attribute strength distribution, such as mean and variance, can be used to estimate the metric.

- Last, invariants in the momentary distribution of the attribute strength can be used to estimate roughly the current metric. Well-known invariants include the gray-world assumption (Hurlbert 1986) and the assumption that the brightest item is perfectly white.

It is highly likely that other ways of calibrating the visual metric exist, including combinations of the above calibration methods. However, another, perhaps more interesting situation occurs when the visual metric is left uncalibrated.

4.4.3 Uncalibrated visual metrics: partial flexibility

The only information available for the accumulation and matching processes mentioned above is the measured attribute as represented by its scale value on the visual metric. When this metric is left uncalibrated, memory standards inevitably become fuzzy due to variability in the scale value as caused by the flexibility of the visual metric. Therefore, besides the difficulties that arise when this scale value is to be compared with memory standards, these memory standards themselves become less precise. It may therefore be useful to restrict the flexibility of the visual metric.

Assume that such restricted, or partial, flexibility can be formalized by introducing the *eternal distribution* $\mu_e(x)$, that is, the momentary distribution $\mu(x)$ integrated over a long-term interval. We can now introduce a weighting parameter $0 \leq \lambda \leq 1$ expressing the relative importances of $\mu(x)$ and $\mu_e(x)$ for the current visual metric $s(x)$. When Eq. (4.23) is generalized to

$$\frac{ds(x)}{dx} = \frac{S}{N}[\lambda\mu(x) + (1 - \lambda)\mu_e(x)], \qquad (4.29)$$

λ can be regarded as the *degree of flexibility* of the visual metric. The influence of flexibility on the variability in the scale value is now easily shown to be proportional to the degree of flexibility, since

$$
\begin{aligned}
s_0 &= \int_{-\infty}^{x_0} \frac{ds(x)}{dx} dx \\
&= \frac{S}{N} \int_{-\infty}^{x_0} [\lambda\mu(x) + (1 - \lambda)\mu_e(x)] dx \\
&= \frac{S}{N} \lambda \int_{-\infty}^{x_0} \mu(x) dx + \frac{S}{N}(1 - \lambda) \int_{-\infty}^{x_0} \mu_e(x) dx, \qquad (4.30)
\end{aligned}
$$

where the first term represents the variability in the scale value s_0 for a given attribute strength x_0. This variability, together with variability in x_0 itself and together with the influences of external and internal noise, will be the main source of fuzziness of memory standards when the latter are assumed to be constructed by accumulation of scale values of past observations. Therefore, although partial flexibility will result in reduced discriminative power, memory standards will be less fuzzy and the comparison of scale values with these standards will be facilitated. The *degree* of flexibility of the visual metric then becomes the subject of optimization in which both discriminability and identifiability play a role.

To conclude, the combination of an uncalibrated flexible metric with rigid memory standards should lead to contextual effects when what is observed is judged in relation to what is represented in memory. Interestingly, Yendrikhovskij, Blommaert & de Ridder (1999b) found such contextual effects. In a series of experiments, subjects were asked to judge the similarity of the object color of a banana located on differently colored backgrounds and displayed on a CRT to what they thought was the color of a prototypical ripe banana. Subjects' judgments showed a significant influence of the color of the background. This influence suggests that what is compared to the memory prototype is the *apparent* object color, that is, the object color as it is observed *without correction for the color of the surroundings*. This is an important experimental result that seems to indicate that visual metrics are *not* calibrated.

4.5   Conclusions

We have shown that flexibility of a metric can be exploited to optimize the overall discriminative power of this metric. We have derived mathematical expressions for optimal metrics, and we have shown that such optimal metrics exhibit properties that correspond to well-known visual phenomena, such as brightness constancy and crispening. We have argued that these phenomena can therefore be understood as logical consequences of visual metrics being flexible and optimal with respect to discriminative power.

Furthermore, we have briefly investigated some of the consequences of flexible visual metrics for the process of visual identification. We have examined the problems that arise when observations represented on a flexible visual metric are compared to memory, and we have proposed two mechanisms for dealing with these problems. The first proposed mechanism is based on calibrating the visual metric. The second proposed mechanism leaves the visual metric uncalibrated and instead restricts the flexibility of the visual metric.

4.A Appendix. Probability of a topological error

Consider the mapping from attribute strength x to scale value s. Ideally, for two items i and j this mapping will be given by $s_i = s(x_i)$ and $s_j = s(x_j)$. However, due to the presence of noise we need a statistical description of this mapping:

$$
\begin{aligned}
s_i &= N\{s(x_i), \sigma[s(x_i)]\} \\
s_j &= N\{s(x_j), \sigma[s(x_j)]\},
\end{aligned}
\tag{4.31}
$$

where N is the normal probability density function, x is the ideal noise-less attribute strength, $s(x)$ the scale function, and $\sigma(s)$ the total noise level. Now, if we assume that item j is the successor of item i in terms of ordering the items by attribute strength, we may use the approximations $x_j - x_i = 1/\mu(x)$, $s(x_j) - s(x_i) = 1/\eta(s)$ and $\sigma[s(x_i)] = \sigma[s(x_j)] = \sigma(s)$, where $\mu(x)$ and $\eta(s)$ are the momentary distributions of attribute strength and associated scale value, respectively. We may now write the probability density function for the scale value difference $ds = s_j - s_i$ as

$$
ds = N[1/\eta(s), \sigma(s)\sqrt{2}].
\tag{4.32}
$$

A topological error occurs when item j becomes the predecessor of item i in terms of ordering the items by scale value, that is, when $ds < 0$. To find the probability of a topological error, we must therefore integrate Eq. (4.32) from $ds = -\infty$ to $ds = 0$:

$$
p_e(s) = \int_{-\infty}^{0} N[1/\eta(s), \sigma(s)\sqrt{2}]d(ds).
\tag{4.33}
$$

Using the linear transformation $u = [ds - 1/\eta(s)]/\sigma(s)$, this can be expressed as

$$
p_e(s) = \int_{-\infty}^{-[\eta(s)\sigma(s)\sqrt{2}]^{-1}} N(0,1)du,
\tag{4.34}
$$

or, equivalently, as

$$
p_e(s) = \frac{1}{2} - \frac{1}{2}\operatorname{erf}\frac{1}{2\eta(s)\sigma(s)}.
\tag{4.35}
$$

Chapter 5

Predicting the Usefulness and Naturalness of Color Reproductions

We present algorithms for predicting the usefulness and naturalness of color reproductions of natural scenes. The algorithms are based on a computational model of the stages that lead to an observer's impression of the usefulness and naturalness of an image. These stages are (1) the perception, or internal quantification, of color; (2) the construction of a memory standard for an object's color based on its color as observed in the past; and (3) matching of observed object colors with memory standards. In the first of the above stages, the internal quantification of color, the concept of (partially) flexible metrics (Chapter 4) plays a central role.

To test the usefulness algorithm, it was used to predict the discriminability of detail in black-and-white images of which the contrast was manipulated by applying an s-shaped transform on CIE 1976 lightness L^*. The naturalness algorithm was tested by using it to predict the naturalness of the grass, skin, or sky areas of images of which the color was manipulated by shifting CIE 1976 hue angle h_{uv} and scaling CIE 1976 saturation s_{uv} of the grass, skin, or sky areas of the images. The predictions produced in these tests correspond quite well to experimentally obtained judgments of human subjects.

This chapter is a slightly modified version of Janssen & Blommaert (2000b), Predicting the usefulness and naturalness of color reproductions, *Journal of Imaging Science and Technology*, vol. 44, no. 2, pp. 93–104. Reprinted with permission of IS&T, The Society for Imaging Science and Technology, Springfield, VA, USA.

5.1 Introduction

Image quality is often considered in terms of a difference signal between the current, reproduced image and its "unprocessed" or "original" version. Well-known examples of this approach include JND (just-noticeable difference) maps (Daly 1993) and SQRI (square root integral) measures (Barten 1990), both of which are based on processing of the image and its original by implementations of visual front-end models and subsequent calculation of a difference measure from the two processed images. A serious drawback of this approach is that the fundamental question of what image quality *is* remains unanswered, as is the related question of what exactly the "original" of an image is. Thus, measures for *differences* between two versions of an image can be calculated; however, what this tells about image quality remains unclear. Fitting the predictions of such models to experimentally obtained quality judgments of human subjects is usually the only means of attempting to make this translation.

We will follow a different approach here that does not suffer from this drawback. Essential to this approach is that we regard the quality of an image as its *adequacy* as an input to visuo-cognitive information processing (Chapter 3). The output of this visuo-cognitive processing determines in turn how well an observer is able to respond to occurrences in the outside world. Thus, in this view, the quality of an image becomes indeed a *meaningful* attribute of an image, telling how well the image can be employed as a source of information about the outside world.

When looking more closely at visuo-cognitive processing of images (see Fig. 5.1), we can discern three processing stages: (1) the construction of an internal representation of the image; (2) the interpretation of this representation by means of matching it with representations stored in memory; and (3) semantic processing of the interpreted scene to formulate a proper response. For these stages to be completed successfully, the image should in general satisfy two main requirements: (1) the internal representation of the image should be precise; and (2) the match between the representation of the image and memory should be close. We refer to the degree to which an image satisfies these two requirements as the *usefulness* and the *naturalness* of the image, respectively.

Evidence for the appropriateness of the above description of image quality in terms of usefulness and naturalness was found in a series of experiments

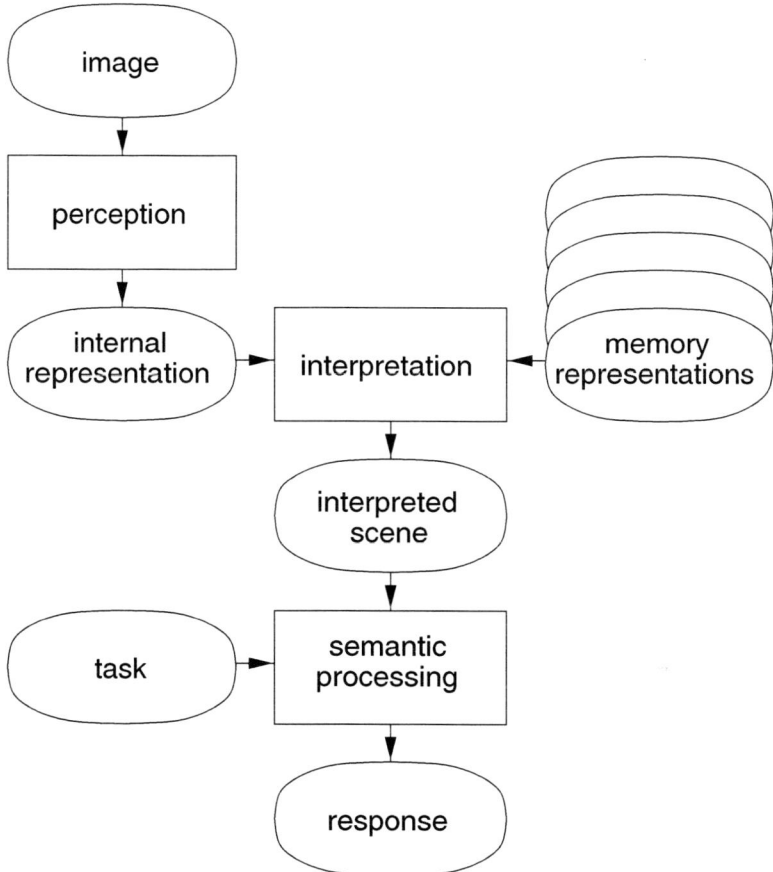

Figure 5.1: Visuo-cognitive processing of images. In this diagram, ellipses denote representations of information, and rectangles denote processes transforming one representation into another.

(see Chapter 3). In these experiments, the separate influences of usefulness and naturalness on image quality were revealed by varying color temperature and chroma of a set of images of natural scenes and asking subjects to judge the quality and naturalness of the manipulated images. Manipulating color temperature was expected to influence only the naturalness of the images, whereas manipulating chroma was expected to influence both naturalness and usefulness. Indeed, quality was found to have a one-to-one relationship to naturalness when color temperature was manipulated, whereas systematic differences between quality and naturalness were found when chroma was manipulated.

To predict the usefulness and naturalness of an image, we need to know what the internal representation of an image looks like. In other words, we need to know which attributes are represented and what *metrics*[1] are used to quantify these attributes. Here, we will simply assume that the structure of the internal representation can be adequately described as a set of measured attributes such as position, shape, size, texture, brightness, and color. When we focus on the attributes brightness and color, as we will do in the remainder of this chapter, we may conclude that the metrics used to quantify these attributes need not be rigid; that is, the scale function need not be constant over time. Moreover, when the scale function of these metrics is allowed to vary in time, this flexibility can be used to improve the discriminative power of the metric (Watt 1989, Watt 1991). Extending this idea, it is possible to find expressions for metrics that are optimal with respect to discriminative power (Chapter 4). Interestingly, such optimal metrics exhibit properties resembling several well-known characteristics of color vision, such as adaptation, crispening, and brightness and color constancy.

Our aim here is to show how the ideas about visuo-cognitive processing of images, combined with the ideas about flexible metrics, can be used to implement algorithms for predicting the usefulness and naturalness of reproductions of color images of natural scenes. The proposed usefulness algorithm is first presented with a set of images from which it calculates the luminance and chromaticity distributions of the pixels of the entire set

---

[1]A metric is a system used for the quantification of measurements. A metric is defined by its *origin, unit, scale,* and *scale function.* Origin and unit together constitute the scale, upon which measurements of the strength of a certain attribute are represented by their *scale values.* The exact relation between attribute strength and corresponding scale value is given by the scale function.

of images, as well as the luminance and chromaticity distributions of the pixels of the image of interest. These distributions are used to calculate the (partially) flexible metrics upon which the brightness and color distributions of the image of interest are represented. The algorithm then calculates overall discriminability from these distributions.

Besides the above steps, the naturalness algorithm also calculates the brightness and color distributions of the areas of each image containing grass, (Caucasian) skin, and sky. Averaged over the entire set of images, the brightness and color distributions of the grass, skin, and sky areas represent the algorithm's "memory standards" for grass, skin, and sky. The algorithm calculates naturalness by comparing the brightness and color distributions of the grass, skin, or sky areas of the image of interest with these memory standards. As we will show, the predictions produced by the usefulness and naturalness algorithms correspond well with experimentally obtained judgments of human subjects.

5.2   Metrics for brightness and color

The concept of (partially) flexible metrics plays a central role in the algorithms we present here, and we will therefore start by introducing this concept. A *metric* is the instrument to quantify the measurement of an attribute's strength. The constituents of a metric are its *scale*, defined by an *origin* and a *unit*, and its *scale function*, which defines the relation between attribute strength and corresponding *scale value* (Watt 1989, Watt 1991). For physical (or biological) implementations of metrics, we may assume that the range of the metric is finite—that is, the scale has fixed lower and upper bounds—and that the precision with which scale values can be represented is limited, for example due to the presence of noise in the system that encodes the scale values. Therefore, when objects are to be discriminated by measurements of their attribute strengths, we find that the discriminative power is essentially determined by the *ratio* of scale value difference to noise level.

The central assumption we will make here is that the metrics used for the quantification of brightness and color are optimal with respect to their discriminative power. Assuming a constant noise level, discriminative power can be increased by increasing scale value differences. One way to accomplish this would be to simply stretch the scale to a larger range. This, however, is no solution here since the range of the scale was assumed to be limited and constant. Another, slightly more complicated way to increase scale value differences is to *locally* stretch the scale at those locations where an increase in discriminative power is desired, by locally increasing the derivative of the scale function, and to compress the scale elsewhere. In this way, the range of the scale can be preserved while scale value differences can be selectively increased. An example of this principle is shown in Fig. 5.2.

Extending this idea, it can be shown that to optimize *overall* discriminative power, the derivative of the scale function should be proportional to the momentary distribution of the attribute strength, as explained in Chapter 4. This result, which is found under the conditions of an internal noise level (due to, for example, random variability in neuronal response rates) that is constant and an external noise level (due to, for example, photon noise) that is negligibly small, can intuitively be understood by realizing that to optimize overall discriminative power, the amount of stretching of the scale should be largest for those ranges of attribute strength that occur most frequently. Such a mechanism bears a close resemblance to the image processing tool known as histogram equalization (Ballard & Brown 1982) and, most importantly, requires the scale function to be *flexible*. Interestingly, such flexible, optimal metrics exhibit properties resembling some well-known visual phenomena, such as dark and light adaptation, crispening (Whittle 1994a, Whittle 1994b), and brightness and color constancy.

The above type of metric is optimal with respect to discriminative power. However, for the aim of identification a metric must satisfy other, partially conflicting requirements. Most notably, identification requires that the scale function be *rigid*, to facilitate the comparison of what is observed at the present with what has been observed in the past. To satisfy this requirement and to simultaneously preserve discriminative power, the degree of flexibility of the metric must be restricted, yet not reduced to zero. We will therefore use the concept of partially flexible metrics (Chapter 4). Mathematically, such metrics can be specified by

$$\frac{ds(x)}{dx} \propto \lambda \mu(x) + (1 - \lambda)\mu_e(x), \tag{5.1}$$

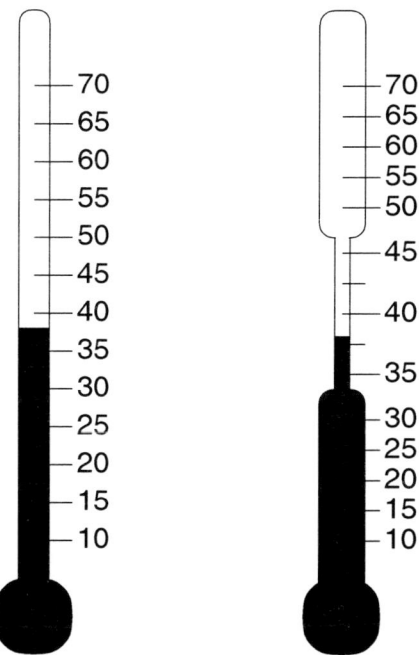

Figure 5.2: An example of how discriminative power can be increased by locally stretching the scale. For the thermometer, the scale function relates the attribute strength "temperature" to the measure "length of a mercury column." The exact relation between temperature and column length is determined by the amount of mercury in the reservoir and the diameter of the glass tube. For the thermometer at the right, discriminative power in the range 35° to 45° has been increased by locally reducing the diameter of the glass tube. The range of the scale, that is, the maximum length of the mercury column, has nevertheless been preserved by increasing the diameter of the glass tube elsewhere.

where x is the attribute strength, $s(x)$ the scale function, $\mu(x)$ the momentary distribution of the attribute strength,[2] $\mu_e(x)$ the "eternal" distribution of the attribute strength, and $0 \leq \lambda \leq 1$ the degree of flexibility of the metric. Here, the eternal distribution $\mu_e(x)$ can be thought of as the momentary distribution $\mu(x)$ integrated over a long-term interval.

Measurement usually involves a stage in which a sensor converts the attribute to be quantified into another attribute that is more accessible for quantification. For example, the thermometer of Fig. 5.2 converts the attribute "temperature" into the attribute "length of a mercury column," which can be quantified easily. The receptors in the retina perform a similar task by converting a complex spectrum of radiated or reflected energy into the set of three values we experience as a color. These three values can be regarded as three separate dimensions, with each having its own metric assigned to it, which make up the attribute color. In the algorithms we present here, we have chosen to use luminance Y and CIE 1976 chromaticity coordinates u' and v' for these dimensions. We have made this choice for two reasons: (1) the dimensions Y, u', and v' resemble the early-visual dimensions of color, namely brightness, red–green, and yellow–blue; and (2) Y, u', and v' coordinates can be calculated relatively easily for digitized images displayed on a CRT.

In practice, to calculate what the metrics for Y, u', and v' look like for a particular image we therefore need to perform the following two steps. First, the momentary and eternal distributions μ and μ_e must be calculated separately for Y, u', and v'. To model the eternal distribution μ_e we calculated the frequency distributions of the Y, u', and v' coordinates of the pixels of a set of 77 images of natural scenes taken from two Kodak photo CDs; and to model the momentary distribution μ we calculated the frequency distributions of the Y, u', and v' coordinates of the pixels of the image of interest. For these calculations we assumed that the images were to be displayed on a PAL (European color television) compliant CRT with γ corrected to the value 2.5. Second, the metrics for Y, u', and v' were calculated from the above distributions for a particular degree of flexibility λ by calculating the weighted sum $\lambda\mu + (1 - \lambda)\mu_e$ and integrating the result. Figure 5.3 summarizes this procedure with an example.

---

[2]In this chapter, $\mu(x)$ is defined such that $\int_X \mu(x)dx = 1$, where X is the range of the attribute strength x. Note that this definition differs from the one used in Chapter 4, where $\mu(x)$ was defined such that $\int_X \mu(x)dx = N$, where N is the number of items.

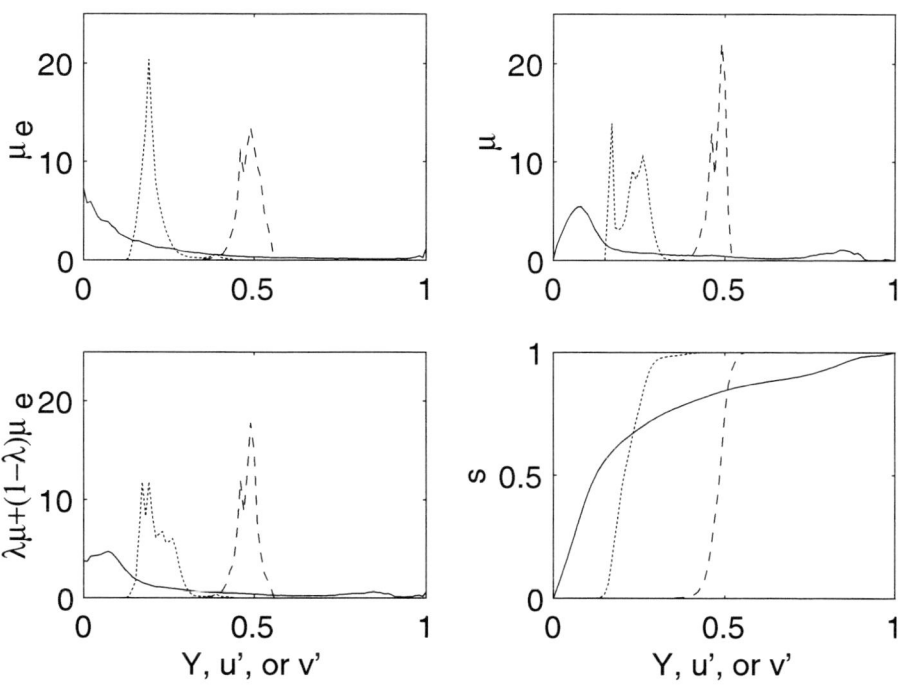

Figure 5.3: A calculated set of (Y, u', v') metrics for an image. The upper left panel shows calculated (Y, u', v') frequency distributions for the entire set of 77 images, representing the eternal distributions. The upper right panel shows calculated (Y, u', v') frequency distributions for one particular image, representing the momentary distributions. The lower left panel shows the weighted sum of momentary and eternal distributions for $\lambda = 0.5$. Integrating these distributions yields the set of metrics for Y, u', and v' for this image. Note that the ranges of u' and v' on the horizontal axes are set to $[0, 1]$ and that the range of Y is normalized to fit in this range. Plots for Y are drawn with solid lines, those for u' with dotted lines, and those for v' with dashed lines.

5.3 Predicting usefulness

In the introduction we defined usefulness in terms of the degree of precision of the internal representation of an image. Here, precision does not necessarily imply that an item's attributes as they are represented internally should correspond one-to-one to the physical characteristics of the item. Instead, the *ability to discriminate* between items in the outside world on the basis of their internally quantified attributes is likely to be a more meaningful criterion for precision. In this section we will therefore define a measure for usefulness that is based on the idea that the usefulness of an image is essentially determined by the overall discriminability of the items in the image. An essential stage in the definition of such a measure is the internal quantification of attribute strength. For this we will use the concept of (partially) flexible metrics presented in the previous section, and limit ourselves to the attributes brightness and color.

5.3.1 Discriminability

In the previous section we have assumed that scale values can only be represented with a limited precision, for example due to the presence of noise in the system encoding these scale values. Discriminability of two items by their measured attribute strengths will therefore be essentially determined by (1) the scale value difference between the items; and (2) the noise level. The measure for discriminability that we will use here is the probability of a topological error. Topological errors occur when the ordering of a set of items by their scale values differs from the ordering of the same set of items by their physical attribute strengths, and the occurrence of topological errors therefore is a strong indicator that discriminability is poor. Assuming Gaussian noise properties, the probability of a topological error, p_{err}, can be shown (see Chapter 4) to be given by

$$p_{err} = \frac{1}{2} - \frac{1}{2} \operatorname{erf} \frac{d}{2\sigma}, \tag{5.2}$$

where d is the ideal, noiseless scale value difference between any pair of items and σ the noise level. In Fig. 5.4, p_{err} is plotted versus d/σ on a logarithmic scale. The figure shows that p_{err} increases asymptotically to a maximum value of 0.5 when d/σ approaches zero, and that p_{err} remains

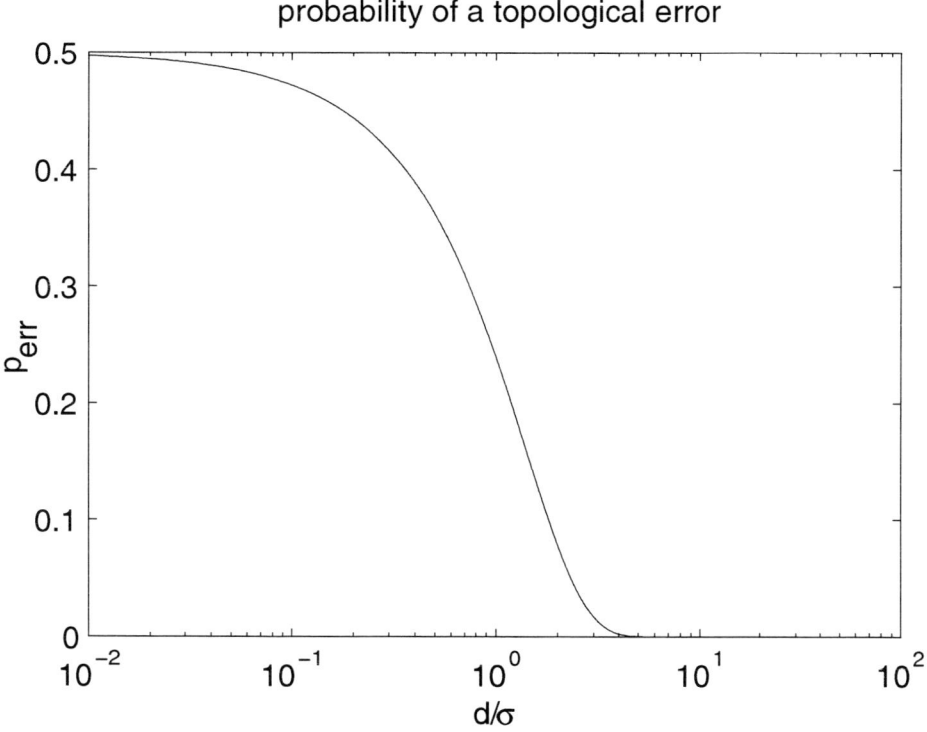

Figure 5.4: The probability of a topological error, p_{err}, versus the ratio of scale value distance d to noise level σ (on a logarithmic scale).

essentially constant when d/σ decreases below 10^{-1} or when d/σ increases beyond 10^{1}.

When the attribute strength distribution of the set of items is given by $\mu(x)$, the associated scale value distribution $\eta(s)$ on the metric $s(x)$ will be given by

$$\eta(s) = \mu(x) \left[\frac{ds(x)}{dx} \right]^{-1}, \qquad (5.3)$$

where $ds(x)/dx$ is the derivative of the scale function, which for a (partially) flexible metric is given by Eq. (5.1). The probability of a topological error will be highest for neighboring items. For neighboring items on the metric, the scale value difference d will in close approximation be given by $1/N\eta(s)$, where N is the number of items in the set. We may therefore approximate Eq. (5.2) by

$$p_{err}(s) = \frac{1}{2} - \frac{1}{2} \, \mathrm{erf} \, \frac{1}{2N\eta(s)\sigma(s)}. \tag{5.4}$$

To obtain the overall probability of a topological error, we must integrate this expression along the entire range S of the metric

$$P_{err} = \int_S p_{err}(s)\eta(s)ds. \tag{5.5}$$

The measure P_{err} must be calculated for each dimension separately. For multidimensional attributes like color, overall discriminability D must somehow be derived from the values found for P_{err} along the n individual dimensions. Since there is no obvious way in which this can be done, we will decrease the set of possible solutions by imposing some desired characteristics. Assuming that overall discriminability D is normalized to the range zero to one, these characteristics are (1) when P_{err} increases for one or more dimensions, then overall discriminability D should decrease; (2) when $P_{err} = 0$ (that is, discriminability is perfect) for at least one dimension, then overall discriminability $D = 1$; and (3) when $P_{err} = 1/2$ (that is, discriminability is poorest) for all n dimensions, then discriminability $D = 0$. Perhaps the simplest way to satisfy these characteristics is when

$$D = 1 - 2^n \prod_{i=1}^{n} P_{err,i}. \tag{5.6}$$

To conclude, the measure we now have was derived for a set of N items. Although for Mondrian-like images the individual patches composing such images may be regarded as the items, for the category of natural images it is unclear what exactly these items are and, therefore, what the value of N is for a particular image. In our calculations we have made the arbitrary choice to set N to the value 100. Furthermore, assuming approximately 100 JNDs (just-noticeable differences) along each dimension, we have set σ to the value 0.01 and S to the range $[0, 1]$. For the ratio d/σ these choices lead to an average value of exactly one, which in Fig. 5.4 lies in the center of the interval where P_{err} is most sensitive to changes in d/σ. In general, although absolute values found for D will depend on the choices made for N, S, and σ, we have found that when ratios of D values are used to compare different images or different versions of one image, results are quite robust to changes in d/σ of up to one order of magnitude from the value we have used here.

In practice, calculation of the usefulness of an image can now be performed as follows. First, the metrics for Y, u', and v' must be calculated for the image following the procedure discussed in the previous section. Next, using Eq. (5.3), the scale value distributions $\eta(s)$ can be calculated from the momentary distributions $\mu(x)$ and the scale functions $s(x)$ for the dimensions Y, u', and v', separately. From the obtained scale value distributions, P_{err} can then be calculated for the individual dimensions using Eqs. (5.4) and (5.5). Finally, overall discriminability D is found by substituting the obtained results in Eq. (5.6). Part of this procedure is summarized in Fig. 5.5 with an example.

5.3.2 Results and discussion

In this subsection we will compare predictions of the above algorithm with experimentally obtained judgments of human subjects. To this end, we manipulated the brightness contrast of four digitized black-and-white images of natural scenes by applying a pixelwise, s-shaped transformation on CIE 1976 lightness L^*:

$$L^{*'} = \left(\frac{L^* - L_{min}^*}{L_{ave}^* - L_{min}^*}\right)^{\gamma} (L_{ave}^* - L_{min}^*) + L_{min}^*$$

$$(\text{for } L_{min}^* \leq L^* \leq L_{ave}^*)$$

$$L^{*'} = \left(\frac{L_{max}^* - L^*}{L_{max}^* - L_{ave}^*}\right)^{\gamma} (L_{ave}^* - L_{max}^*) + L_{max}^*$$

$$(\text{for } L_{ave}^* < L^* \leq L_{max}^*), \tag{5.7}$$

where L^* represents the original lightness of a pixel, $L^{*'}$ the new value, and where L_{min}^*, L_{max}^*, and L_{ave}^* represent the minimum, maximum, and average

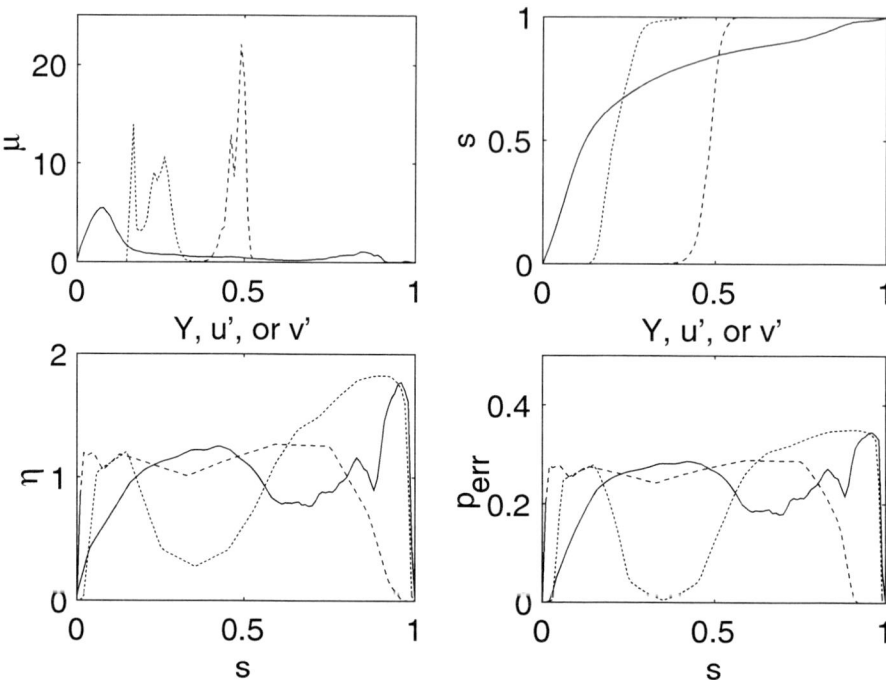

Figure 5.5: Summary of the procedure to obtain the probability of a topological error, P_{err}, for an image. The upper panels show the momentary distributions μ (left) and scale functions s for Y, u' and v' (right) for the image of interest (note that we again used $\lambda = 0.50$; these plots are identical to the ones shown in Fig. 5.3 in the upper right and lower right panels). The lower left panel shows the associated scale value distributions η, and the lower right panel shows p_{err} for these distributions, assuming that N is 100 and σ is 0.01. Integrating the areas under the curves yields P_{err}. In this particular case P_{err} is 0.23, 0.20, and 0.23 for Y, u' and v', respectively. Plots for Y are drawn with solid lines, those for u' with dotted lines, and those for v' with dashed lines.

lightness of the pixels in the original image, respectively. Furthermore, the parameter γ is specified in terms of a gain factor g as

$$\gamma = 10^g. \tag{5.8}$$

The transformation is shown in Fig. 5.6. Applying this transformation will for $g < 0$ decrease the brightness contrast, and for $g > 0$ increase the brightness contrast of the image. Minimum and maximum lightness of the image remain at their original values, while the average lightness value in the image remains at approximately the same level. We used nine versions of each scene, with gain factor values of -0.60, -0.45, -0.30, -0.15, 0, 0.15, 0.30, 0.45, and 0.60.

Eight subjects were instructed to judge overall visibility of detail of the resulting images, which were shown in random order, with three replications, on a PAL-compliant CRT. For their judgments, subjects used an 11-point numerical scale ranging from 0 ("bad") to 10 ("excellent"). Results, averaged over subjects and replications, are shown as the solid curves in Fig. 5.7. Furthermore, results produced by the algorithm are shown as dotted, short-dashed, and long-dashed curves for $\lambda = 0$, $\lambda = 0.25$, and $\lambda = 0.50$, respectively. Note that to facilitate a comparison, all curves in the figure were z-scored.

Figure 5.7 shows that subjects' judgments of visibility of detail are slightly asymmetrical around $g = 0$, with visibility of detail being maximal for $g = 0$ (images 1 and 4) to $g = 0.15$ (images 2 and 3). For predictions made by the algorithm this asymmetry is more pronounced, with maximum discriminability occurring for $g = 0$ (image 4), $g = 0.15$ (image 1), and $g = 0.30$ (images 2 and 3). In general, predictions made by the algorithm correspond well to subjects' judgments of visibility of detail for values of g below zero. For values of g above zero, the algorithm tends to overestimate discriminability compared to subjects' judgments. Subjects report that decreased visibility of detail for g smaller than zero is due primarily to decreased contrast of the manipulated images. Likewise, for values of g slightly larger than zero, subjects report increased visibility of detail due to increased contrast. For higher values of g, however, subjects report that visibility of detail is decreased, due to the existence of areas in the manipulated images where detail is lost due to clipping to either black or white. Apparently, the algorithm is underestimating the impact of this

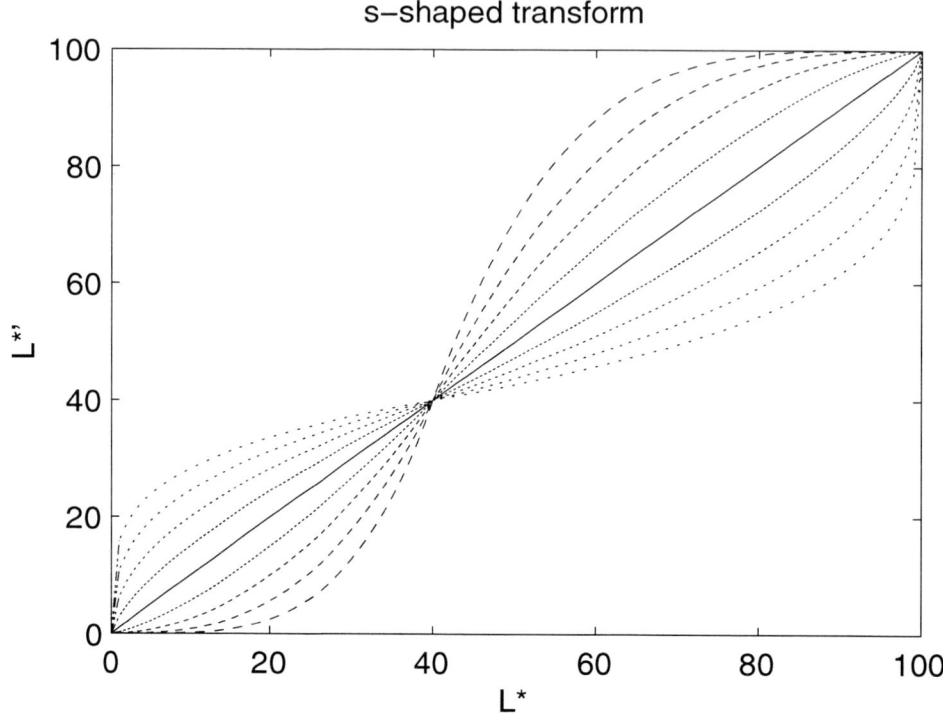

Figure 5.6: The s-shaped transformation we applied to CIE 1976 lightness L^*. On the horizontal axis the original lightness, and on the vertical axis the lightness after applying the s-shaped transformation. Curves shown are for $g = -0.60$, $g = -0.45$, $g = -0.30$, and $g = -0.15$ (dotted curves, with decreasing dot gap for increasing g), $g = 0$ (solid curve), $g = 0.15$, $g = 0.30$, $g = 0.45$, and $g = 0.60$ (dashed curves, with increasing dash length for increasing g). Minimum, average, and maximum values for L^* are 0, 40, and 100 in this example.

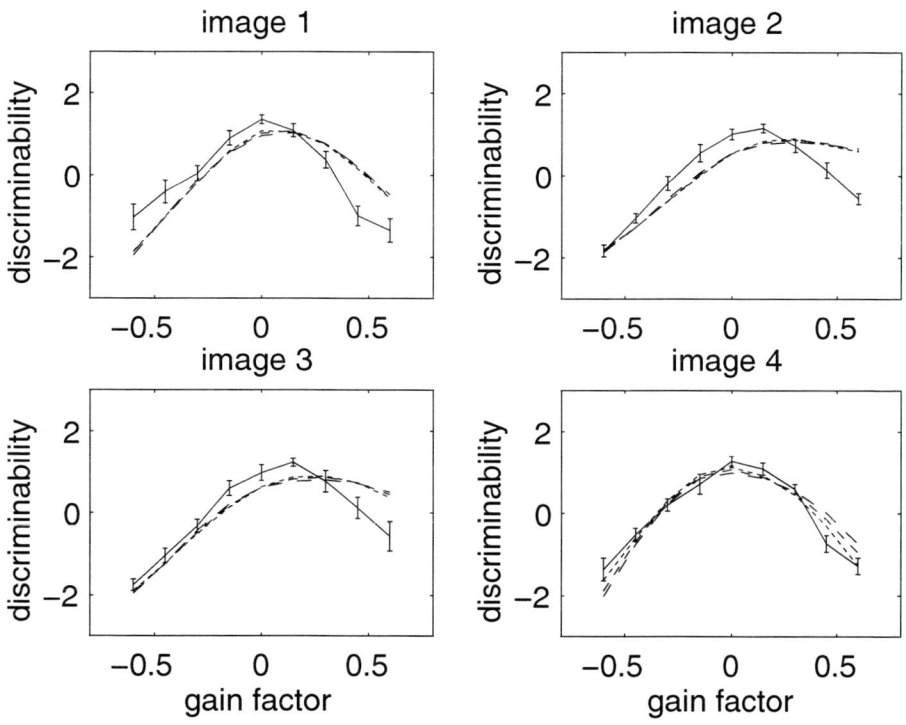

Figure 5.7: Visibility of detail as judged by human subjects (solid lines, the error bars denote a distance of two standard errors in the mean) and overall discriminability D (short-dashed lines for $\lambda = 0$, medium-dashed lines for $\lambda = 0.25$, and long-dashed lines for $\lambda = 0.5$) versus the gain factor of the s-shaped transformation on lightness. Results have been z-scored to facilitate a comparison.

effect on overall visibility of detail. Such underestimation is probably due to the fact that the algorithm is analyzing luminance statistics only globally, that is, without taking into account how these statistics vary from one location in the image to another. Alternatively, subjects' attention may be drawn to areas where clipping occurs, resulting in an overproportional influence of clipping on their judgments. Nevertheless, overall correspondence between model predictions and subjects' judgments is quite good, certainly given the assumptions and simplifications that underlie the algorithm.

5.4   Predicting naturalness

In the introduction we defined naturalness in terms of the degree of match between the internal representation of an image and memory. Realizing that the content of images is usually made up of objects that are more or less familiar, we may specify this further as the degree to which perceived object attributes match remembered object attributes, or *memory standards*. As stated in the introduction, in the algorithm we present here we will restrict ourselves to the attributes color and brightness. To predict naturalness, we therefore need to specify (1) how colors are internally quantified; (2) how the memory standard for a particular object's color is constructed; and (3) how the perceived color of an object is matched with the memory standard for that object's color. We will address the second and third issue here, since the first issue has already been addressed in the section about metrics for brightness and color.

5.4.1   The construction of memory standards

When observing a particular scene, the scale function of a partially flexible metric will be determined by the distribution of the attribute strength for that scene (the momentary distribution) and by the distribution of the attribute strength for all scenes observed in the past (the eternal distribution). Observation of a particular object in this scene will therefore result in a scale value for that object that is determined by, first, the attribute strength for that object and, second, the scale function at the moment of observation. For a given attribute strength x and scale function $s(x)$, the corresponding scale value s is given simply by $s = s(x)$. However, the attribute strength measured for a particular object at a particular moment will usually be given by a distribution of values instead of one unique value. For example, the color of grass in a particular scene is not uniform but instead varies from location to location. This attribute strength distribution will have its associated scale value distribution on the metric. The relation between the attribute strength distribution $\mu_o(x)$ and the associated scale value distribution $\eta_o(s)$ is simple and given by

$$\eta_o(s) = \mu_o(x) \left[\frac{ds(x)}{dx} \right]^{-1}, \tag{5.9}$$

where $ds(x)/dx$ is given by Eq. (5.1). The scale value distribution η_o for an individual object will therefore depend not only on the attribute strength distribution of the object itself but also on the attribute strength distribution of the scene in which it is located.

We will assume that the memory standard for an object is constructed by the accumulation of past observations of that object; see Fig. 5.8. As explained above, the observation of an object will result in a distribution of scale values on the metric. Scale value distributions of past observations therefore constitute the information from which memory standards are constructed. How this accumulation is performed is unclear. Here, we will simply assume that the memory standard for a particular object is constructed by a long-term integration of the scale value distributions observed for this object. Alternatively, accumulation may also be performed by calculation and storage in memory of parameters such as mean and standard deviation of observed scale value distributions. We have chosen not to pursue this approach to avoid making assumptions about the shapes of the observed distributions.

Figure 5.9 shows calculated memory standards for the objects grass, skin, and sky. These standards were obtained by calculating the Y, u', and v' distributions of the image areas containing grass, skin, and sky. The associated scale value distributions were then calculated using Eq. (5.3). Finally, the memory standards were constructed by averaging the obtained scale value distributions over the sets of images containing grass, skin, or sky. This procedure was repeated for three degrees of flexibility: $\lambda = 0$, $\lambda = 0.25$, and $\lambda = 0.5$.

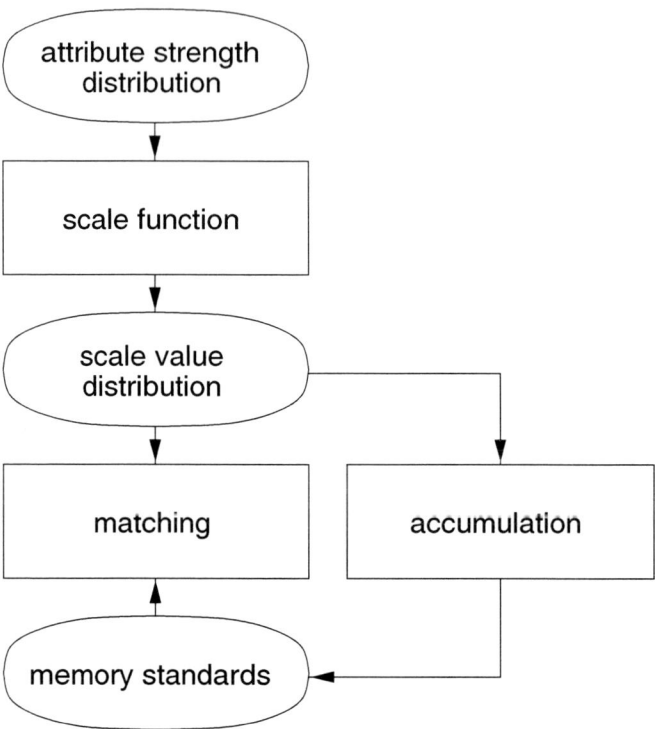

Figure 5.8: A simple model for the accumulation and matching processes. Matching is performed by means of comparing scale value distributions with standards stored in memory. The memory standards themselves are assumed to be constructed by means of accumulation of scale value distributions of past observations.

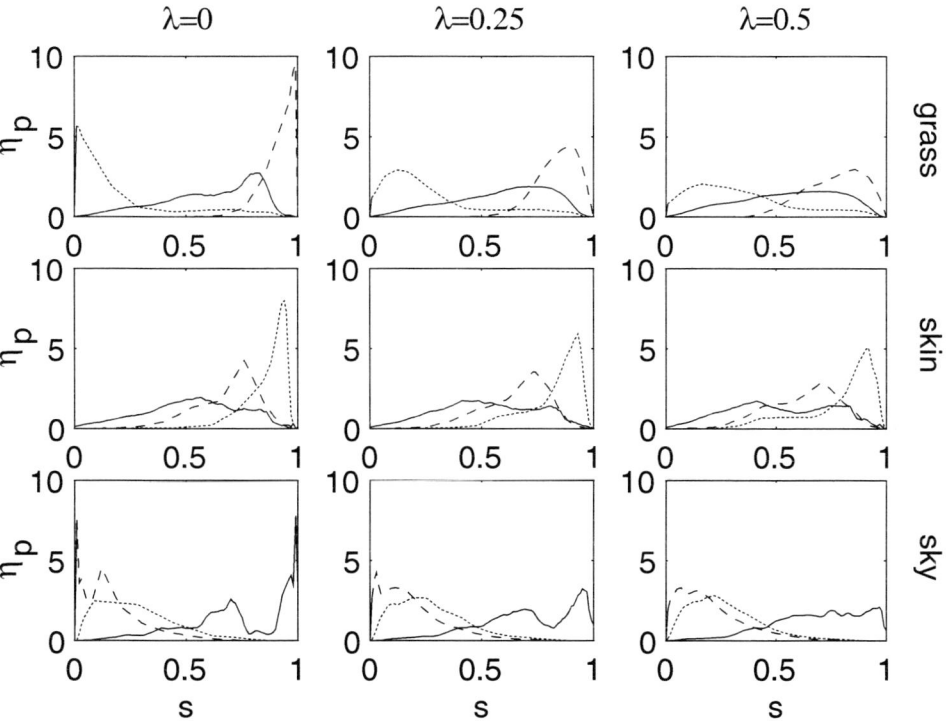

Figure 5.9: Calculated memory standards $\eta_p(s)$ for grass (upper row), skin (middle row), and sky (lower row) for degree of flexibility 0 (left column), 0.25 (middle column), and 0.5 (right column). Plots for Y are drawn with solid lines, those for u' with dotted lines, and those for v' with dashed lines.

5.4.2 Matching perceived object colors with memory standards

The aim of the matching process is to identify an object using its observed scale value distribution. The obvious way in which this task can be accomplished is by matching the observed scale value distribution with memory standards and selecting the memory standard for which the degree of match is highest. The performance of such a mechanism may be judged by three main criteria: (1) *success*; that is, the overall probability of identifying an object correctly; (2) *sensitivity*; that is, the degree to which distinctions can be made in the identification of objects; and (3) *robustness*; that is, the degree to which the identification of an object remains stable under small variations in the observed scale value distribution as caused by, for example, noise.

There is no unique solution to this problem, and for the algorithm we present here we have made the rather arbitrary choice to express the degree of match between an observed scale value distribution and a memory standard by a normalized correlation measure:

$$m[\eta_o(s), \eta_p(s)] = \frac{\int \eta_o(s)\eta_p(s)ds}{\sqrt{\int \eta_o^2(s)ds \int \eta_p^2(s)ds}}, \tag{5.10}$$

where $m(.)$ is the degree of match, η_o the observed scale value distribution, and $\eta_p(s)$ the memory standard. The main advantage of this measure is its robustness, specifically its independence of assumptions about the shape of the distributions that are to be matched. It produces results that lie in the range zero (perfect mismatch) to one [perfect match, when $\eta_o(s) = \eta_p(s)$]. However, since we have to deal with three dimensions, a measure of overall match must be derived from the degrees of match along the individual dimensions. The criteria that may be imposed on such a measure are (1) overall match should increase monotonically with the degree of match in the individual dimensions; (2) overall match should be zero when there is at least one dimension for which the degree of match is zero; and (3) overall match should be one only when the degree of match is one for all individual dimensions. The obvious candidate for this measure is a simple product of the degrees of match along the individual dimensions:

$$m = m_Y \cdot m_{u'} \cdot m_{v'}, \tag{5.11}$$

where m is the overall degree of match, and where m_Y, $m_{u'}$, and $m_{v'}$ are the degree of match along the dimensions Y, u', and v', respectively.

To conclude, we have defined naturalness as the degree of match between perceived object attributes and memory standards. Assuming that an object has already been identified by selection of the memory standard for which the degree of match is highest, this degree of match represents the naturalness of that object. With m we have a measure for predicting the naturalness of the grass, skin, and sky areas of color reproductions of natural scenes. To predict the naturalness of the entire image, the above predictions should in principle be calculated for each individual object depicted in the image. However, since most color manipulations applied to images are global, that is, not restricted to specific locations in the image, a weighted average of naturalness predictions for a limited set of objects depicted in the image will usually be sufficient to predict the naturalness of the entire image. Yendrikhovskij et al. (1999a) have shown that such an approach can indeed be used successfully to predict the naturalness of an entire image from naturalness predictions for the grass, skin, and sky areas of this image.

5.4.3   Results and discussion

In this subsection we compare naturalness predictions produced by the above algorithm with experimental results reported by Yendrikhovskij et al. (1999a). In these experiments, human subjects had to judge the naturalness of manipulated color reproductions of natural scenes that were displayed on a CRT. To this end, the portions of the images showing grass, skin, or sky were manipulated by shifting CIE 1976 hue angle h_{uv} and by scaling CIE 1976 saturation s_{uv}. The obtained naturalness judgments, averaged over subjects and scenes, were then plotted versus the average u' and v' coordinates of the manipulated grass, skin, or sky areas, and subsequently a two-dimensional Gaussian $f(u', v')$ was fitted to this data:

$$f(u', v') \quad \propto \quad \exp - \frac{1}{2(1 - \rho_{u'v'}^2)} (u_n'^2 - 2\rho_{u'v'} u_n' v_n' + v_n'^2)$$

$$u_n' \quad = \quad \frac{u' - \mu_{u'}}{\sigma_{u'}}$$

$$v_n' \quad = \quad \frac{v' - \mu_{v'}}{\sigma_{v'}}, \tag{5.12}$$

where $\mu_{u'}$, $\sigma_{u'}$, $\mu_{v'}$, $\sigma_{v'}$, and $\rho_{u'v'}$ were the free parameters to be fitted. Finally, the authors plotted what they referred to as "one-sigma ellipses" in the (u', v') plane. These ellipses, which connect locations of equal naturalness in the (u', v') plane, are given by the equation

$$-\frac{1}{2(1 - \rho_{u'v'}^2)} (u_n'^2 - 2\rho_{u'v'} u_n' v_n' + v_n'^2) = -\frac{1}{2}. \tag{5.13}$$

To compare the naturalness predictions as produced by the algorithm with the data of Yendrikhovskij et al., we applied the same manipulations to the grass, skin, or sky areas of twelve images that were selected from the set of 77 images.[3] In particular, hue angle h_{uv} was shifted by $-\frac{4}{8}\pi$, $-\frac{3}{8}\pi$, $-\frac{2}{8}\pi$, $-\frac{1}{8}\pi$, 0, $\frac{1}{8}\pi$, $\frac{2}{8}\pi$, $\frac{3}{8}\pi$, and $\frac{4}{8}\pi$, and saturation s_{uv} was scaled by 0.41, 0.51, 0.64, 0.80, 1.00, 1.25, 1.56, 1.95, and 2.44. Naturalness of the manipulated grass, skin, or sky areas was then predicted using the algorithm, and a Gaussian fit was made for each image separately. Resulting one-sigma ellipses for these fits are shown in Fig. 5.10 for four images of which the grass areas were manipulated, in Fig. 5.11 for four images of which the skin areas were manipulated, and in Fig. 5.12 for four images of which the sky areas were manipulated.

In general, locations, sizes, and orientations of the one-sigma ellipses fitted to the algorithm's predictions correspond well to the one-sigma ellipses found by Yendrikhovskij et al. A degree of flexibility $\lambda = 0.25$ produces best results for grass, $\lambda = 0$ produces best results for skin, and $\lambda = 0.50$ produces best results for sky. This seems to indicate that some degree of flexibility is needed to produce best overall results. It would nevertheless be too strong to conclude this, since the validity of such a conclusion is strongly influenced by the characteristics of the set of 77 images that were used to "train" the algorithm. This influence is difficult to estimate since there is no set of images that can be called representative for what humans observe in everyday life. Nevertheless, given the constraints imposed by having to use a

---

[3]We are aware that by selecting our test images from the set of images used to construct memory standards, we are introducing some unwanted correlation between the test images and the memory standards. For the construction of one-sigma ellipses we expect this effect to be negligibly small, since the one-sigma ellipses are constructed using 81 versions (nine saturation settings times nine hue settings) of each test image. Of these 81 versions, only one version corresponds to the image that is originally used in the construction of the memory standard.

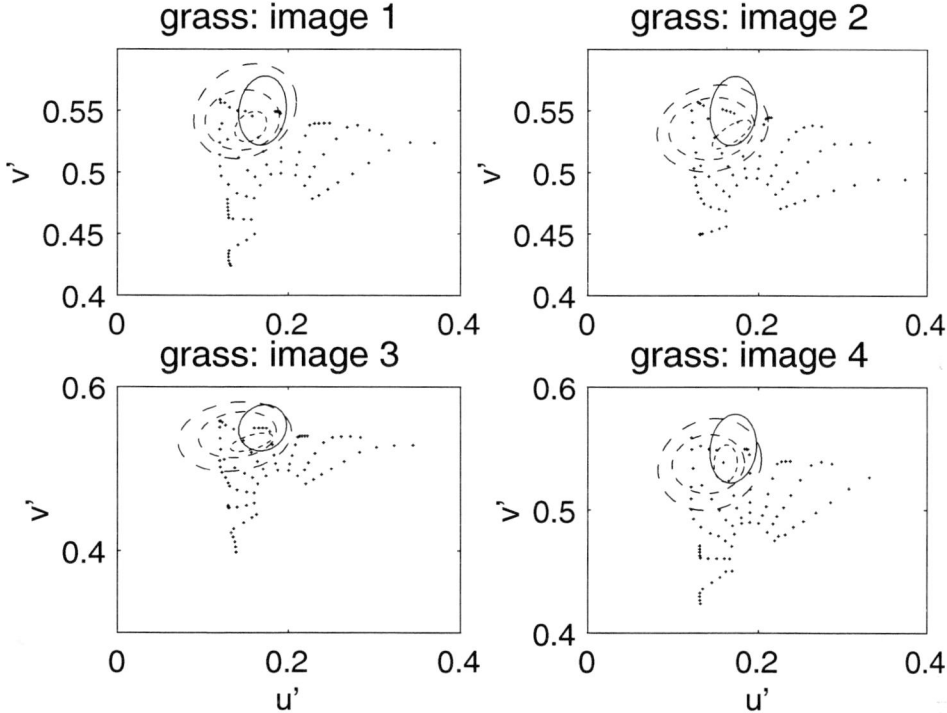

Figure 5.10: Comparison of the results of Yendrikhovskij et al. with predictions made by the algorithm for images of which the grass areas were manipulated. One-sigma ellipses for degrees of flexibility $\lambda = 0$, $\lambda = 0.25$, and $\lambda = 0.50$ are shown with short-dashed, medium-dashed, and long-dashed lines, respectively. The one-sigma ellipse found by Yendrikhovskij et al. is shown with a solid line. The dots represent the average (u', v') coordinates of the grass area of the manipulated images.

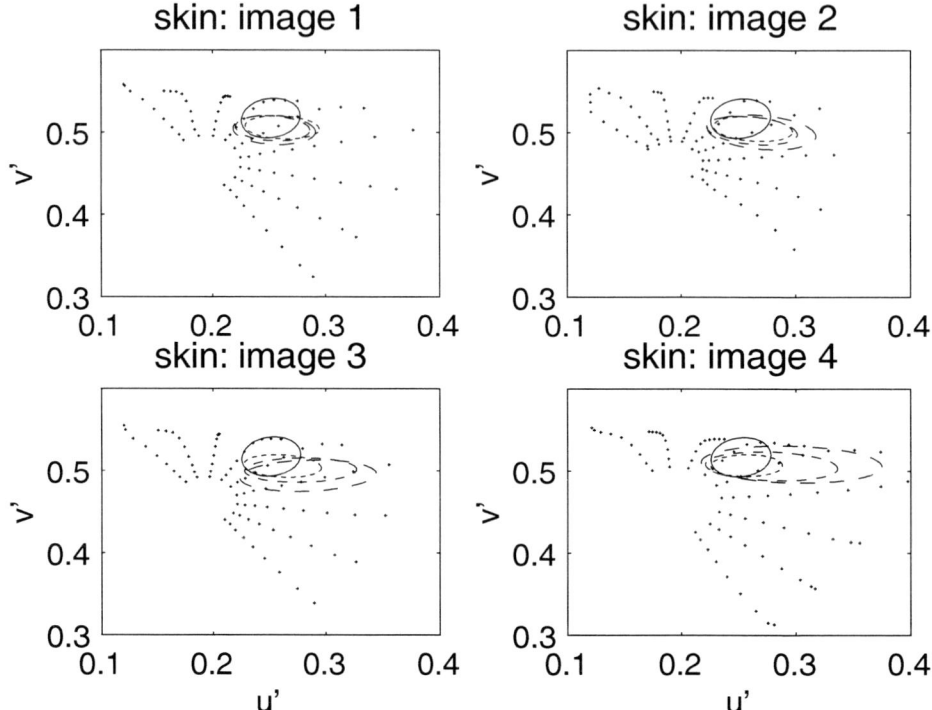

Figure 5.11: Comparison of the results of Yendrikhovskij et al. with predictions made by the algorithm for images of which the skin areas were manipulated. One-sigma ellipses for degrees of flexibility $\lambda = 0$, $\lambda = 0.25$, and $\lambda = 0.50$ are shown with short-dashed, medium-dashed, and long-dashed lines, respectively. The one-sigma ellipse found by Yendrikhovskij et al. is shown with a solid line. The dots represent the average (u', v') coordinates of the skin area of the manipulated images.

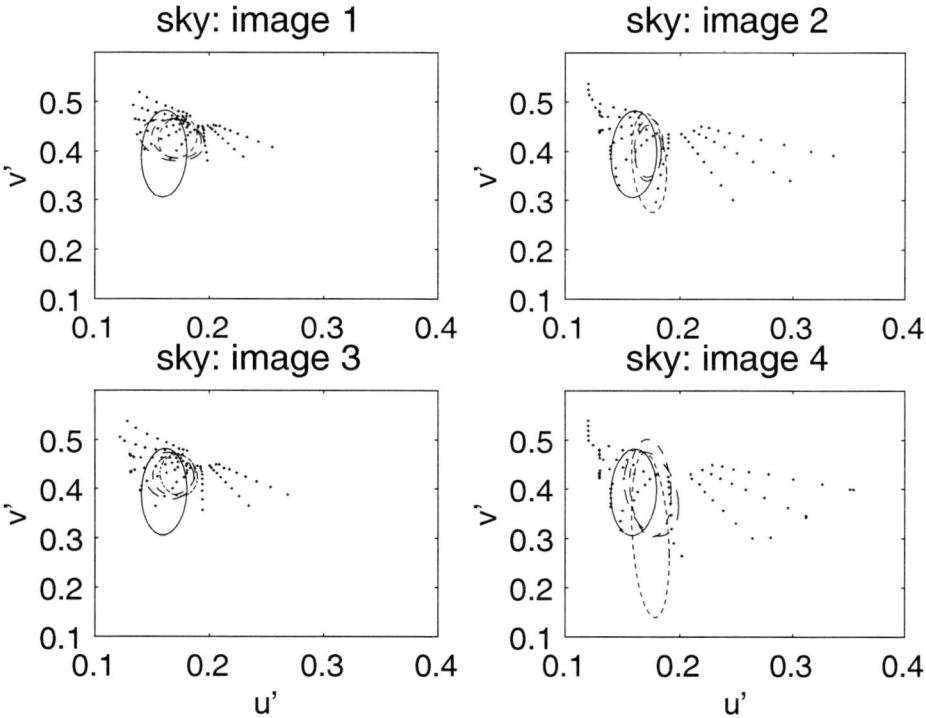

Figure 5.12: Comparison of the results of Yendrikhovskij et al. with predictions made by the algorithm for images of which the sky areas were manipulated. One-sigma ellipses for degrees of flexibility $\lambda = 0$, $\lambda = 0.25$, and $\lambda = 0.50$ are shown with short-dashed, medium-dashed, and long-dashed lines, respectively. The one-sigma ellipse found by Yendrikhovskij et al. is shown with a solid line. The dots represent the average (u', v') coordinates of the sky area of the manipulated images.

limited set of images, the algorithm produces results that correspond quite well with results produced by human subjects.

5.5 Conclusions

We have presented algorithms for predicting the usefulness and naturalness of color reproductions of natural scenes, which are the two principal components determining the quality of these reproductions. The algorithm for predicting usefulness is based on the idea that usefulness is given by the overall discriminability of the items in the image using their observed attributes. The algorithm for predicting naturalness is based on the idea that naturalness is given by the degree of match between object attributes as observed in the reproduction with standards for these attributes as stored in memory. To predict usefulness and naturalness, the following stages must therefore be specified: (1) the observation of object attributes, that is, the internal quantification of these attributes; (2) the construction of memory standards from object attributes as observed in the past; and (3) the matching of observed object attributes with memory standards. For the internal quantification of object attributes we have used the concept of partially flexible metrics presented in Chapter 4; for the specification of memory standards we have assumed a simple accumulation of object attribute distributions observed in the past; and for the matching process we have assumed a simple correlation between observed object attribute distributions and memory standards. The usefulness algorithm was tested by using it to predict the discriminability of detail in manipulated black-and-white images of natural scenes, and the naturalness algorithm was tested by using it to predict the naturalness of the grass, skin, and sky areas of manipulated color reproductions of natural scenes. The predictions produced by the algorithms correspond quite well with experimentally obtained judgments of human subjects.

Chapter 6

Image Quality Revisited

We present a concept for image quality that is based on a definition of quality in terms of the degree to which something satisfies the requirements imposed on it. An answer to the question of what image quality is must therefore necessarily include answers to these questions: what are images? what are images used for? and what are the requirements which the use of images imposes on them? In this chapter we therefore start by formulating answers to these questions. To this end, we distinguish two main requirements that are imposed upon images. First, the items in the image should be successfully discriminable; and, second, the items in the image should be successfully identifiable. Based on the concept of (partially) flexible metrics presented in Chapter 4, we then formulate algorithms for predicting discriminability, identifiability, and overall performance. To demonstrate the validity of this concept, we compare predictions made with these algorithms with experimentally obtained judgments of human subjects.

6.1   Introduction

Image quality is usually described in terms of the presence of visible distortions of the image such as color shifts, blur, noise, or blockiness. Historically, the most common way to model and predict image quality therefore

---

This chapter is a slightly modified version of Janssen & Blommaert (2000a), A computational approach to image quality, *Displays*, vol. 21, no. 4, pp. 129–142. Reprinted with permission of Elsevier Science, Oxford, UK.

has been a quantification of the visibility of these distortions, for example using models of early visual processing or using judgments of human subjects obtained in psychophysical experiments. Since this approach seems so obvious, the question of what image quality really is has traditionally been neglected, denying a better interpretation and deeper understanding of what has been measured and modeled during the past.

Our aim here is to answer the above question of what image quality is, and to present a concept for image quality that is both generic and directly applicable. The concept is generic in the sense that the approach we use here can be used equally well to describe sound or speech quality. Furthermore, it is directly applicable in the sense that the description we give of image quality, and the algorithms we base upon this description, follow straightforwardly from the answer to the above question. Finally, to show the value of this concept for real applications, we show some predictions made with algorithms based on this concept together with corresponding judgments of human subjects.

6.2 What is image quality?

There can be no good answer to the question "What is image quality?" when the question "What is quality?" has not been answered. Perhaps the most general answer to this last question is "the degree to which something satisfies the requirements imposed on it." Though vague, this answer certainly is intuitively correct. We tend to think of things as "bad," "good," or "excellent," according to the degree to which they exhibit desired characteristics (a book; a house), or according to the degree to which they are adequate for the task we want them to perform (a car; a washing machine). So, if we wish to answer the question of what image quality is, we will need to answer the questions (1) what are images? (2) what are images used for? and (3) what are the requirements which the use of images imposes on them?

To begin by answering the first two questions, there can be no doubt that images are the carriers of visual information about the outside world, and that they are used as input to human visual perception. Visual perception itself is part of the three processes perception, cognition, and action, which together constitute human interaction with the environment (see Fig. 6.1). Images, therefore, can be regarded as input to the perception stage of interaction. If we use a rather technical view of perception, we may define perception as the stage of human interaction in which attributes of items in the outside world are measured and internally quantified. The aim of this quantification is essentially twofold. First, items in the outside world can be discriminated from one another using their internally quantified attributes. The result of this process is an essential step toward the construction of higher-level descriptions of scene geometry and object location, descriptions upon which later processes such as navigation in the scene are based. Second, items in the outside world can be identified by comparing their internally quantified attributes with quantified attributes, stored in memory, of similar items observed in the past. Identification of what is depicted in the image is an essential step in the interpretation of scene content; it determines our semantic awareness of what is in the scene.

So far, what we have found is that images are used as input to the perception stage of human interaction, and that the primary task of perception is to measure and internally quantify attributes of items in the outside world with the aim to discriminate and identify these items. These observations lead us to the answer to the third question: the requirements that the use of images—that is, their use as input to perception—imposes on them is that first, the items depicted in the image should be successfully discriminable; and, second, the items depicted in the image should be successfully identifiable. We are now able to formulate the following answer to the question of what image quality is: (1) The quality of an image is the adequacy of this image as input to visual perception. (2) The adequacy of an image as input to visual perception is given by the discriminability and identifiability of the items depicted in the image.

6.3   The internal quantification of attributes

If we wish to predict image quality, we will need to specify how discrimination and identification of items is performed, and to do this we need to specify how item attributes are internally quantified. The instrument for the quantification of attribute strength is the *metric*. Metrics are defined by an *origin* and a *unit*, which together constitute the *scale*, and a *scale function*,

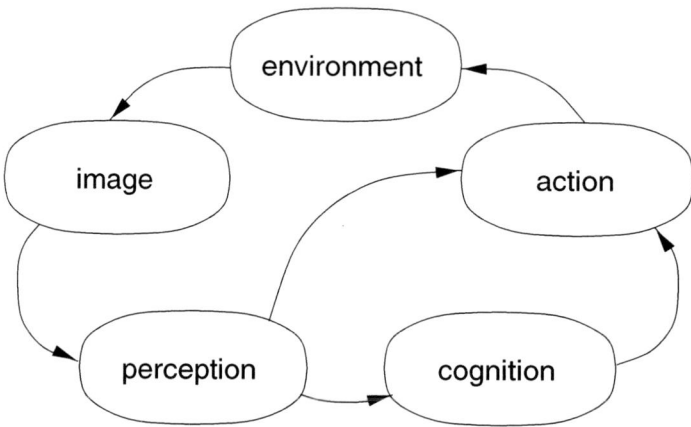

Figure 6.1: Schematic overview of the interaction process. Images are the carriers of information about the environment, and serve as input to visual perception. The result of visual processing is used as input to cognition (for tasks requiring interpretation of scene content) or as input to action (for example in navigation, where the link between perception and action is mostly direct). Since action will in general result in a changed status of the environment, the nature of the interaction process is cyclic.

which relates physical attribute strength to the position on the scale, or *scale value* (Watt 1989, Watt 1991). The scale function is usually assumed to be rigid, that is, constant in time. Essentially all metrics defined by humans are of this type, the reason for this being that we wish to use these metrics to uniquely specify attribute strength in terms of scale value; 1.78 meters should be one and the same length, wherever and whenever it is measured.

As we have pointed out, perception is the stage in human interaction with the environment in which attributes of items in the outside world are measured and internally quantified, with the aim to discriminate and identify these items. The metrics used for this task must somehow be physiologically implemented. Any such physiological implementation, and in general any physical implementation of a metric, will inevitably result in first, a limited scale range; and, second, a limited precision with which scale values can be represented. Therefore, if we consider the limited precision with which scale values can be represented to be a source of noise, the ability to discriminate two items using their internally represented scale values will essentially be determined by the ratio of the difference of their scale values to the noise level. This, together with the limited scale range, represents an upper limit to the discriminative power of any physically implemented metric.

Watt (1991) has argued that the discriminative power of a metric can be increased considerably when the scale function of that metric is allowed to be flexible instead of rigid. Flexibility allows for an adaptation of the scale function to what is currently being measured, thus sacrificing a unique specification of measured item attributes for an improved ability to discriminate these items. In Chapter 4 we have extended this idea to find expressions for scale functions that are optimal with respect to discriminative power. We found that such metrics exhibit properties that resemble several well-known characteristics of human color vision, such as adaptation, crispening, and brightness and color constancy. However, since flexibility of the scale function no longer allows for a unique specification of measured item attributes, the ability to compare what is observed at the moment with what has been observed in the past is reduced, resulting in a reduced ability to identify. The solution for this problem is to use partially flexible scale functions (Chapter 4). The degree of flexibility of such scale functions can then be used to optimize the overall performance in terms of both discriminability and identifiability.

Since the aim of quantification in perception is to discriminate and identify items, and since discrimination and identification are of such importance to a successful interaction, it is to be expected that the properties of the metrics used for this quantification are optimally—or nearly optimally—chosen. We will therefore adopt the concept of partially flexible metrics here, and start by deriving expressions for metrics that are optimal with respect to either discriminability, identifiability, or a combination of these two. Based on the expressions we derive, we will then formulate expressions for discriminability and identifiability. To this end, we will regard the outside world as a simple set of items of which the values of a one-dimensional attribute are measured and quantified with the aim to discriminate and identify these items. We will further simplify the situation by assuming that the influence of external noise sources; that is, noise sources acting directly on the attribute strength, may be neglected. In the last part of this chapter we will apply the obtained measures for discriminability and identifiability to predict the quality of reproduced black-and-white images of natural scenes of which the contrast has been manipulated, and compare the predictions with experimentally obtained judgments of human subjects.

6.4 An optimal metric for overall discriminability

Assume that, at a certain point in time, the strength of an attribute x is measured for a set of N items. The momentary distribution[1] of the strength of x for the items in the set is given by $\mu(x,t)$, where $\mu(x,t)$ is scaled to yield unity when integrated over the range X of the attribute strength:

$$\int_X \mu(x,t)dx = 1. \tag{6.1}$$

The measurement results are represented by scale values on an internal scale s. The range of the scale s is given by S, and the scale function $s(x,t)$ that relates attribute strength x to scale value s is monotonic with x; that is, $ds(x,t)/dx \geq 0$. Furthermore, assume that the precision with which scale

---

[1]We will assume that the momentary distribution $\mu(x,t)$ varies slowly with respect to the time in which adaptationlike effects occur. We assume this since we are not interested in a description of temporal aspects here. At any moment in time we will therefore regard the momentary distribution $\mu(x,t)$ as quasi-static.

values can be represented is limited by the presence of Gaussian noise with mean zero and spread σ. Finally, assume that the influence of external noise on the attribute strength x may be neglected.

The questions we now ask are (1) When the items are to be discriminated using their scale values, what is the overall discriminability of the items in the set (expressed in the overall probability of a topological error) for a given scale function $s(x,t)$? and (2) What scale function $s(x,t)$ will optimize this overall discriminability (that is, minimize the overall probability of a topological error[2])?

6.4.1 A measure for overall discriminability

The scale function $s(x,t)$ maps attribute strength x to scale value s. When the attribute strength distribution of the set of N items is given by $\mu(x,t)$, then the corresponding scale value distribution $\eta[s(x,t)]$ will be determined by the relation

$$\eta[s(x,t)]ds(x,t) = \mu(x,t)dx, \tag{6.2}$$

so that

$$\eta[s(x,t)] = \mu(x,t) \left[\frac{ds(x,t)}{dx}\right]^{-1}. \tag{6.3}$$

When the items are ordered by their attribute strengths, we find that the attribute strength differences between subsequent items are in close approximation given by $\Delta x = 1/N\mu(x,t)$. The corresponding scale value difference will be given by $\Delta s = 1/N\eta[s(x,t)]$, with $\eta[s(x,t)]$ given in Eq. (6.3). However, due to the presence of noise on the internal scale values, the corresponding scale value s for a certain attribute strength x will be given by a Normal probability density function with mean $s(x,t)$ and spread σ. For an item with attribute strength x and its successor with attribute strength $x + \Delta x$, we therefore find for the scale value *difference* a Normal probability

---

[2]Topological errors occur when the ordering of the set of items by their scale values differs from the ordering by their attribute strengths. The occurrence of topological errors is an adequate indicator of poor discriminability, and we will therefore substitute the problem of minimizing the probability of a topological error for the problem of maximizing discriminability.

density function with mean $\Delta s = 1/N\eta[s(x,t)]$ and spread $\sigma\sqrt{2}$. A topological error occurs when $\Delta s < 0$. To find the probability of a topological error, we thus have to integrate the above probability density function from $\Delta s = -\infty$ to $\Delta s = 0$:

$$p_d(x,t) = \int_{-\infty}^{0} N\{\Delta s; 1/N\eta[s(x,t)], \sigma\sqrt{2}\}d(\Delta s), \qquad (6.4)$$

which after substitution of $u = \{\Delta s - 1/N\eta[s(x,t)]\}/\sigma\sqrt{2}$ can be written as

$$p_d(x,t) = \int_{-\infty}^{-1/N\sigma\sqrt{2}\eta[s(x,t)]} N(u; 0, 1)du, \qquad (6.5)$$

or

$$p_d(x,t) = \frac{1}{2} - \frac{1}{2}\text{erf}\,\frac{1}{2N\sigma\eta[s(x,t)]}. \qquad (6.6)$$

The overall probability of a topological error is now given by

$$P_d(t) = \int_X p_d(x,t)\mu(x,t)dx. \qquad (6.7)$$

Substituting Eqs. (6.6) and (6.3) in this, we obtain

$$P_d(t) = \int_X \left[\frac{1}{2} - \frac{1}{2}\text{erf}\,\frac{ds(x,t)/dx}{2N\sigma\mu(x,t)}\right]\mu(x,t)dx. \qquad (6.8)$$

6.4.2 Optimizing overall discriminability

The question we ask here is: what scale function $s(x,t)$ will minimize the overall probability of a topological error $P_d(t)$ for a given attribute strength distribution $\mu(x,t)$? Referring to Eq. (6.8), we have to solve

$$\min_{ds(x,t)/dx} \int_X \left[\frac{1}{2} - \frac{1}{2}\text{erf}\,\frac{ds(x,t)/dx}{2N\sigma\mu(x,t)}\right]\mu(x,t)dx. \qquad (6.9)$$

In Chapter 4 we have shown that the solution to this problem is given by

$$\frac{ds(x,t)}{dx} = S\mu(x,t), \tag{6.10}$$

which shows that the optimal scale function $s(x,t)$ depends on the momentary distribution of the attribute strength and therefore should be flexible. For an optimal metric, the ideal, noiseless scale value s for a particular attribute strength x is therefore given by

$$s = s(x,t) = S \int_{-\infty}^{x} \mu(u,t)du. \tag{6.11}$$

Substituting the above solution in Eq. (6.8), the overall probability of a topological error for an optimal metric is given by

$$P_{d,min} = \frac{1}{2} - \frac{1}{2}\mathrm{erf}\frac{S}{2N\sigma}, \tag{6.12}$$

which defines the upper limit of discriminability for given values of S, σ, and N. In contrast, the lower limit of discriminability is given by the maximum value of $P_d(t)$, which independently of the values of S, σ, and N is always given by

$$P_{d,max} = \frac{1}{2}. \tag{6.13}$$

Discriminability may now be specified on a scale from zero (worst) to one (best) by using

$$D(t) = 1 - \frac{P_d(t) - P_{d,min}}{P_{d,max} - P_{d,min}}. \tag{6.14}$$

6.5 An optimal metric for overall identifiability

We will assume that identification of an item is performed by means of selecting the item whose accumulated distribution of scale values as measured in the past is most likely to give rise to the presently measured scale value. Furthermore, we will assume that an optimal scale function as specified in Eq. (6.11) is used for the mapping of attribute strength x to internal scale value s.

The questions we wish to answer here, are (1) What do the accumulated distributions of measured scale values for the items look like? (2) In what decision rule does the above identification procedure result? (3) What is the overall identifiability at a certain point in time (expressed in the overall probability of an identification error) using the above identification procedure? and (4) What scale function will optimize this overall identifiability (that is, minimize the overall probability of an identification error)?

6.5.1 The accumulation of scale value distributions

We have found that, for a metric that is optimal with respect to discriminability, the scale value s for a particular attribute strength x at a particular point in time t is given by Eq. (6.11). Over time, the average scale value $\bar{s}$ measured for a particular item with attribute strength x will therefore be given by

$$
\begin{aligned}
\bar{s} = s(x) &= \frac{1}{T} \int_T s(x,t) dt \\
&= \frac{1}{T} \int_T S \int_{-\infty}^{x} \mu(u,t)(du)(dt) \\
&= S \int_{-\infty}^{x} \frac{1}{T} \int_T \mu(u,t)(dt)(du) \\
&= S \int_{-\infty}^{x} \mu(u) du,
\end{aligned} \tag{6.15}
$$

where $\mu(x)$ is the average over time of the momentary attribute strength distribution $\mu(x,t)$:

$$
\mu(x) = \frac{1}{T} \int_T \mu(x,t) dt. \tag{6.16}
$$

At any particular point in time, the deviation $\Delta s(x,t)$ from the average scale value given by Eq. (6.15) will be given by

$$\Delta s(x,t) \;=\; s(x,t) - s(x)$$

$$= \; S \int_{-\infty}^{x} \mu(u,t)du - S \int_{-\infty}^{x} \mu(u)du$$

$$= \; S \int_{-\infty}^{x} [\mu(u,t) - \mu(u)]du. \tag{6.17}$$

Assuming that the above difference $\Delta s(x,t)$ may be adequately described by a Normal probability density function, the mean of this distribution will be zero and the spread $\Delta s(x)$ will be given by

$$\Delta s(x) \;=\; \sqrt{\frac{1}{T}\int_{T}[\Delta s(x,t)]^2 dt}$$

$$= \; \sqrt{\frac{1}{T}\int_{T}\left\{ S \int_{-\infty}^{x} [\mu(u,t) - \mu(u)]du \right\}^2 dt}. \tag{6.18}$$

Note that, due to the separate and independent influence of noise on the scale values, the total spread $\hat{s}$ will be higher and given by

$$\hat{s} = \sqrt{[\Delta s(x)]^2 + \sigma^2}. \tag{6.19}$$

The accumulated scale value distribution for an item with attribute strength x may now be characterized by a mean $\bar{s}$ given by Eq. (6.15), and a spread $\hat{s}$ around this mean given by Eq. (6.19). Equation (6.19) shows that the spread of the distributions will not be constant along the scale s. However, for the remainder we will assume that the spread varies slowly along s and that, locally on s, the spread may be considered constant.

6.5.2   A decision rule for identification

We have assumed that identification is performed by selecting the item whose accumulated distribution of scale values as measured in the past is most likely to give rise to the presently measured scale value. Furthermore, we have found that the accumulated scale value distribution for an item with attribute strength x can be characterized by a mean and a spread given by Eqs. (6.15) and (6.19), respectively. Since we have assumed that, locally on s, the spread of the scale value distributions is constant, selection of the most likely distribution reduces to the simple task of selection of the distribution whose mean lies closest to the currently measured scale value. Such a strategy results in the construction of decision bounds along the scale s that lie midway between the distribution means of adjacent distributions. Our task here is to find an expression for the positions of these decision bounds.

We will assume here that a set of N_i items, representative for the entire set of observed items, is to be identified. Specifically, we will assume that (1) the momentary attribute strength distribution of the items to be identified may at any point in time be approximated by the attribute strength distribution of the set of N observed items; and (2) the attribute strength distribution of the entire set of N_i items to be identified may be approximated by the average over time of the momentary attribute strength distribution of the set of observed items, as specified by Eq. (6.16).

Having assumed this, we are now able to formulate an expression for the positions of the decision bounds for the set of N_i items defined above. We have found that the scale value distribution mean $\bar{s}$ of an item with attribute strength x is given by $s(x)$. When the attribute strength distribution of the set of N_i items is given by $\mu(x)$, the resulting distribution $\eta[s(x)]$ of the scale value distribution means will be given by

$$\eta[s(x)]ds(x) = \mu(x)dx, \tag{6.20}$$

so that

$$\eta[s(x)] = \mu(x)\left[\frac{ds(x)}{dx}\right]^{-1}. \tag{6.21}$$

From Eq. (6.15) it follows that

$$\frac{ds(x)}{dx} = S\mu(x). \tag{6.22}$$

For the distribution of the scale value distribution means we therefore find

$$\eta[s(x)] = 1/S; \tag{6.23}$$

that is, the distribution means of the N_i items are distributed homogeneously along the scale s and consequently are a distance S/N_i apart. For an item with attribute strength x this means that the decision bounds for selecting this item are given by $s(x) \pm S/2N_i$, with $s(x)$ given by Eq. (6.15).

6.5.3 A measure for overall identifiability

We are now ready to formulate an expression for the overall identifiability at a particular point in time. The scale value observed for an item with attribute strength x will, at a particular point in time, be given by a Normal probability density function with mean $s(x, t)$ given by Eq. (6.11) and a spread around this mean of σ. The item will be correctly identified when this scale value lies within the decision bounds that identify it as the item with attribute strength x; that is, when it lies within the interval $s(x) \pm S/2N_i$. To find the probability of an identification error, we therefore have to integrate this probability density function over the interval $s = s(x) - S/2N_i$ to $s = s(x) + S/2N_i$ and subtract the result from one:

$$p_i(x, t) = 1 - \int_{s(x)-S/2N_i}^{s(x)+S/2N_i} N[s; s(x, t), \sigma] ds, \tag{6.24}$$

which after substitution of $u = [s - s(x, t)]/\sigma$ and of $\Delta s(x, t) = s(x, t) - s(x)$ can be written as

$$p_i(x, t) = 1 - \int_{[-\Delta s(x,t)-S/2N_i]/\sigma}^{[-\Delta s(x,t)+S/2N_i]/\sigma} N(u; 0, 1) du, \tag{6.25}$$

or as

$$p_i(x, t) = 1 + \frac{1}{2}\text{erf}\frac{\Delta s(x, t) - S/2N_i}{\sigma\sqrt{2}} - \frac{1}{2}\text{erf}\frac{\Delta s(x, t) + S/2N_i}{\sigma\sqrt{2}}. \tag{6.26}$$

The overall probability of an identification error at a particular point in time is now given by

$$P_i(t) = \int_X p_i(x,t)\mu(x,t)dx, \tag{6.27}$$

which again assumes that the attribute strength distribution of the items to be identified may at any point in time be approximated by the attribute strength distribution of the set of N items. Substituting Eq. (6.26), we obtain

$$P_i(t) = \int_X \left[1 + \frac{1}{2}\text{erf}\,\frac{\Delta s(x,t) - S/2N_i}{\sigma\sqrt{2}} - \frac{1}{2}\text{erf}\,\frac{\Delta s(x,t) + S/2N_i}{\sigma\sqrt{2}} \right] \mu(x,t)dx, \tag{6.28}$$

where $\Delta s(x,t)$ is given by Eq. (6.17).

6.5.4 Optimizing overall identifiability

The problem we wish to solve here, is: what scale function $s(x,t)$ will minimize the overall probability of an identification error $P_i(t)$ for given attribute strength distributions $\mu(x,t)$ and $\mu(x)$? Referring to Eq. (6.28), we have to solve

$$\min_{\Delta s(x,t)} \int_X \left[1 + \frac{1}{2}\text{erf}\,\frac{\Delta s(x,t) - S/2N_i}{\sigma\sqrt{2}} - \frac{1}{2}\text{erf}\,\frac{\Delta s(x,t) + S/2N_i}{\sigma\sqrt{2}} \right] \mu(x,t)dx, \tag{6.29}$$

and then derive $s(x,t)$ from $\Delta s(x,t)$ and $s(x)$ using Eqs. (6.15) and (6.17). The solution to this problem is easily found: if we wish to minimize Eq. (6.28), we will have to find a $\Delta s(x,t)$ such that $p_i(x,t)$ is minimized for all x. Referring to Eq. (6.26), we are looking for a $\Delta s(x,t)$ such that

$$\text{erf}\,\frac{\Delta s(x,t) + S/2N_i}{\sigma\sqrt{2}} - \text{erf}\,\frac{\Delta s(x,t) - S/2N_i}{\sigma\sqrt{2}} \tag{6.30}$$

is *maximized*. This equation represents the area under the curve of a Normal probability density function with mean zero and spread σ over the interval $\Delta s(x,t) - S/2N_i$ to $\Delta s(x,t) + S/2N_i$. Evidently, this area is maximized when the interval lies symmetrically around the mean of the distribution; that is, when $\Delta s(x,t) = 0$. From Eq. (6.17) it follows that for $\Delta s(x,t)$ to be zero, $s(x,t)$ should be equal to $s(x)$, with $s(x)$ given by Eq. (6.15). For an optimal metric, we therefore find here:

$$s(x,t) = s(x) = S \int_{-\infty}^{x} \mu(u)du, \tag{6.31}$$

with $\mu(u)$ the average over time of the momentary frequency distribution $\mu(x,t)$. Equation (6.31) shows that the scale function $s(x,t)$ should in this case be rigid. Substituting the solution we find here in Eq. (6.28), the overall probability of an identification error for an optimal metric is given by

$$P_{i,min} = 1 - \text{erf}\, \frac{S}{2\sqrt{2}N_i\sigma}, \tag{6.32}$$

which defines the upper limit of identifiability for given S, σ, and N_i. The lower limit of identifiability will be given by the maximum value of $P_i(t)$, which independently of S and σ and for $N_i > 1$ is given by

$$P_{i,max} = 1. \tag{6.33}$$

Identifiability may now be specified on a scale from zero (worst) to one (best) by using

$$I(t) = 1 - \frac{P_i(t) - P_{i,min}}{P_{i,max} - P_{i,min}}. \tag{6.34}$$

6.6 Estimating the number of discriminable and identifiable items

To estimate the number of discriminable items, we return to Eq. (6.12). When we rewrite this equation to

$$N = \frac{S}{2\sigma\,\text{erf}^{-1}\,(1 - 2P_d)}, \tag{6.35}$$

this equation allows us to estimate the maximum number of discriminable items using an optimal metric for a given allowable error probability P_d of, say, 25%, and a given internal dynamic range S/σ. A plot of the maximum number of discriminable items versus the dynamic range S/σ is shown as the solid line in the left panel of Fig. 6.2.

In a similar way, we may rewrite Eq. (6.32) to

$$N_i = \frac{S}{2\sqrt{2}\sigma \operatorname{erf}^{-1}(1 - P_i)}. \tag{6.36}$$

Here, too, the maximum number of identifiable items using an optimal metric may be estimated for a given allowable error probability P_i and a given dynamic range S/σ. The solid line in the right panel of Fig. 6.2 shows a plot of the maximum number of identifiable items versus the dynamic range S/σ obtained this way for $P_i = 25\%$.

Figure 6.2 shows that, for an allowable error probability of 25%, the number of discriminable items N using a metric that is optimal with respect to discriminability is about the same as the dynamic range S/σ. The number of identifiable items N_i using a metric that is optimal with respect to identifiability is, however, a factor 2.4 smaller than the dynamic range.

6.7 Partial flexibility

We have found expressions for overall discriminability [Eq. (6.8)] and overall identifiability [Eq. (6.28)]. Furthermore, we have found that optimizing overall discriminability requires that the scale function be flexible [Eq. (6.11)], and that optimizing overall identifiability requires that it be rigid [Eq. (6.31)]. What we wish to do here is to find a *partially flexible* scale function that is chosen such that the total performance (in terms of both discriminability and identifiability) is optimized. This problem can be solved using regularization; see, for example, Poggio, Torre & Koch (1990). If we express the total performance $T(t)$ in terms of the discriminability $D(t)$ and identifiability $I(t)$, total performance can be optimized by solving

$$\min_{s(x,t)} T(t) = \epsilon D(t) + (1 - \epsilon)I(t), \tag{6.37}$$

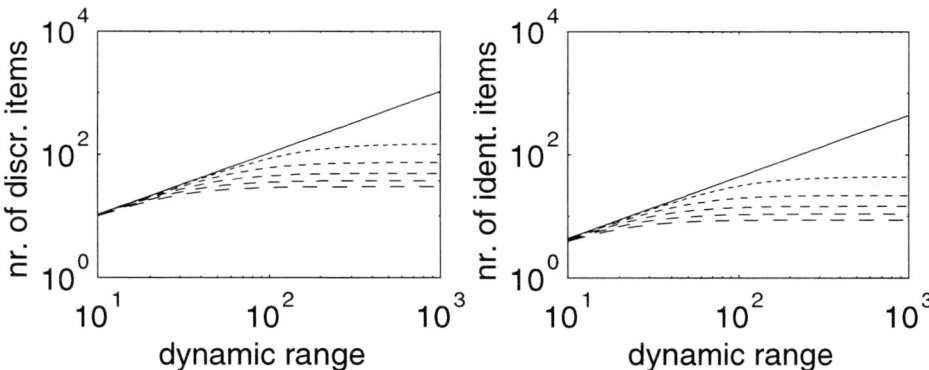

Figure 6.2: The number of discriminable items (left panel) and identifiable items (right panel) versus the dynamic range S/σ of the scale, for a maximum allowable error probability of 25%. The solid lines represent optimum performance; that is, a completely flexible metric for discriminability and a completely rigid metric for identifiability. The dashed lines show the influence of decreased flexibility for the number of discriminable items, and the influence of decreased rigidity for the number of identifiable items. See section 6.7.3 for how the dashed curves have been obtained.

where ϵ expresses the relative importances of discriminability and identifiability for the overall performance.

Here, we will assume that the general solution to this problem belongs to the class of scale functions:

$$s(x,t) = \lambda S \int_{-\infty}^{x} \mu(u,t)du + (1-\lambda)S \int_{-\infty}^{x} \mu(u)du, \qquad (6.38)$$

where the optimal value of the parameter λ will depend on the choice made for ϵ. The parameter λ can be interpreted as the *degree of flexibility* of the scale function, since for $\lambda = 0$ we obtain a completely rigid scale function that is optimal with respect to identifiability, while for $\lambda = 1$ we obtain a completely flexible scale function that is optimal with respect to discriminability. Note that the average over time of a partially flexible metric as specified in Eq. (6.38) is still given by

$$
\begin{aligned}
\bar{s} &= \frac{1}{T} \int_{T} s(x,t)dt \\
&= \frac{1}{T} \int_{T} \left[\lambda S \int_{-\infty}^{x} \mu(u,t)du + (1-\lambda)S \int_{-\infty}^{x} \mu(u)du \right] dt \\
&= \lambda S \int_{-\infty}^{x} \frac{1}{T} \int_{T} \mu(u,t)(dt)(du) + (1-\lambda)S \int_{-\infty}^{x} \frac{1}{T} \int_{T} \mu(u)(dt)(du) \\
&= \lambda S \int_{-\infty}^{x} \mu(u)du + (1-\lambda)S \int_{-\infty}^{x} \mu(u)du \\
&= S \int_{-\infty}^{x} \mu(u)du \\
&= s(x). \qquad (6.39)
\end{aligned}
$$

Therefore, the results derived for the accumulation of scale value distributions and the decision rule based upon this accumulation are still valid.

In the remainder we will derive expressions for discriminability $D(t,\lambda)$ and identifiability $I(t,\lambda)$, given a certain degree of flexibility λ. The performance optimization problem as formulated above now becomes

$$\min_{\lambda} \epsilon D(t,\lambda) + (1-\epsilon)I(t,\lambda), \qquad (6.40)$$

which can be solved by taking the derivative with respect to λ and setting the result to zero:

$$\frac{\partial}{\partial \lambda}[\epsilon D(t,\lambda) + (1-\epsilon)I(t,\lambda)] = 0. \tag{6.41}$$

6.7.1 Discriminability for partial flexibility

When substituting Eq. (6.38) into Eq. (6.8), we obtain for the overall probability of a topological error

$$P_d(t,\lambda) = \int_X \left[\frac{1}{2} - \frac{1}{2}\text{erf}\,\frac{\lambda S\mu(x,t) + (1-\lambda)S\mu(x)}{2N\sigma\mu(x,t)}\right]\mu(x,t)dx. \tag{6.42}$$

Discriminability $D(t,\lambda)$ can be derived from this by using Eq. (6.14).

6.7.2 Identifiability for partial flexibility

For the overall probability of an identification error we have already found Eq. (6.28):

$$P_i(t,\lambda) = \int_X \left[1 + \frac{1}{2}\text{erf}\,\frac{\Delta s(x,t) - S/2N_i}{\sigma\sqrt{2}} - \frac{1}{2}\text{erf}\,\frac{\Delta s(x,t) + S/2N_i}{\sigma\sqrt{2}}\right]\mu(x,t)dx. \tag{6.43}$$

For a partially flexible metric as specified by Eq. (6.38), $\Delta s(x,t)$ will be given by

$$\Delta s(x,t) = \lambda S \int_{-\infty}^{x} [\mu(u,t) - \mu(u)]du. \tag{6.44}$$

Identifiability $I(t,\lambda)$ can be derived from this by using Eq. (6.34).

6.7.3 Re-estimating the number of discriminable and identifiable items

Our aim here is to re-estimate the number of discriminable and identifiable items, this time for a partially flexible metric as specified by Eq. (6.38). We will first regard the number of discriminable items. For items having a small difference δx in their attribute strengths, the corresponding scale value difference δs is given by $\delta s = [ds(x,t)/dx]\delta x$, where $ds(x,t)/dx$ for a flexible metric is given by $ds(x,t)/dx = \lambda S\mu(x,t) + (1-\lambda)S\mu(x)$. Furthermore, when the items are ordered by attribute strength, the attribute strength difference between subsequent items is in good approximation given by $\delta x(x,t) = 1/N\mu(x,t)$. For the corresponding scale value difference $\delta s(x,t)$ we therefore find

$$\delta s(x,t) = \frac{1}{N\mu(x,t)}[\lambda S\mu(x,t) + (1-\lambda)S\mu(x)]. \tag{6.45}$$

Defining $\Delta\mu(x,t) = \mu(x,t) - \mu(x)$ we may rewrite this to

$$
\begin{aligned}
\delta s(x,t) &= \frac{1}{N\mu(x,t)}\{\lambda S\mu(x,t) + (1-\lambda)S[\mu(x,t) - \Delta\mu(x,t)]\}. \\
&= \frac{S}{N} - \frac{S}{N}(1-\lambda)\frac{\Delta\mu(x,t)}{\mu(x,t)} \\
&= \overline{\delta s(x,t)} + \widehat{\delta s(x,t)}. \tag{6.46}
\end{aligned}
$$

The second term of this equation may be regarded as a source of variability on the average scale value difference $\overline{\delta s(x,t)} = S/N$, which is proportional to the degree of rigidity $(1-\lambda)$. Since this variability is caused primarily by the term $\Delta\mu(x,t)/\mu(x,t)$, we may not expect this variability to be Normally distributed. If we nevertheless assume a Normal probability density function, the mean of this distribution will be zero and the spread given by

$$\widehat{\delta s} = \sqrt{\frac{1}{X}\int_X \frac{1}{T}\int_T [\widehat{\delta s(x,t)}]^2 (dt)(dx)}. \tag{6.47}$$

For the scale value differences we have already found that the variability as caused by internal noise is given by a Normal probability density function with a spread $\sigma\sqrt{2}$. When we include the above variability as caused by rigidity, and when we assume that the approximation by a Normal probability density function is more or less valid, we find a total equivalent variability that is given by a spread of

$$\sigma' = \sqrt{(\sigma\sqrt{2})^2 + \widehat{\delta s}^2}. \tag{6.48}$$

Substituting σ' for $\sigma\sqrt{2}$ in Eq. (6.35) now allows for an estimation of the number of discriminable items for a partially flexible metric. Resulting plots for $\widehat{\delta s} = \sigma, 2\sigma, \ldots, 5\sigma$ are shown as the dashed lines in the left panel of Fig. 6.2. The figure indeed shows that the number of discriminable items is reduced when the metric is less than completely flexible. However, since $\widehat{\delta s}$ is effectively determined by the product of rigidity $(1 - \lambda)$ and the difference between momentary and average distribution $\Delta\mu(x, t)$, the reduced number of discriminable items is the result of a combined effect, and reduced flexibility will therefore only result in a reduced number of discriminable items when $\mu(x, t)$ is not equal to $\mu(x)$.

Related to this, it is interesting to notice that if we regard the number of *discriminable steps* in attribute strength x, as measured in terms of JNDs (just-noticeable differences), we find that this number should for rigid metrics be essentially equal to the dynamic range S/σ. For a (partially) flexible metric this no longer holds, and the dynamic range S/σ becomes the *lower limit* for the number of discriminable steps. The explanation for this is as follows. For a small step Δx in attribute strength, we find a corresponding scale value difference $\Delta s = [ds(x)/dx]\Delta x$. Any scale value difference Δs that exceeds the noise level σ on the scale effectively corresponds to a discriminable step. Therefore, $\Delta x = \sigma/[ds(x)/dx]$ defines the size of a discriminable step. Due to adaptation of the (partially) flexible metric to the momentary frequency distribution of the attribute strength, $ds(x)/dx$ will always be relatively high for the value of x currently used as the reference for Δx. The resulting discriminable step size Δx will therefore for a (partially) flexible metric be relatively small, and the total number of discriminable steps correspondingly high. The conclusion that may be drawn from this is that measurement of the dynamic range of a certain metric in terms of the number of JNDs will result in an overestimation of the dynamic range when the metric under consideration is (partially) flexible.

We now turn to estimating the number of identifiable items. The use of a partially flexible metric as given in Eq. (6.38) will result in a mapping from attribute strength x to scale value s given by

$$s = s(x,t) = s(x) + \Delta s(x,t), \tag{6.49}$$

with $\Delta s(x,t)$ given by Eq. (6.44). If we assume that $\Delta s(x,t)$ may be approximated by a random variable that is Normally distributed, the spread of the Normal probability density function will be given by

$$\hat{s} = \sqrt{\frac{1}{X}\int_X \frac{1}{T}\int_T [\Delta s(x,t)]^2 (dt)(dx)}. \tag{6.50}$$

Using this approximation, we have found a source of variability that adds to the internal noise on the scale values and that is proportional to the degree of flexibility λ. The total equivalent noise on the scale values is therefore given by a Normal probability density function with mean zero and spread σ' given by

$$\sigma' = \sqrt{\sigma^2 + \hat{s}^2}. \tag{6.51}$$

Substituting σ' for σ in Eq. (6.36), we may now re-estimate the number of identifiable items when a partially flexible metric is used. Resulting plots for $\hat{s} = \sigma, 2\sigma, \ldots, 5\sigma$ are shown as the dashed lines in the right panel of Fig. 6.2.

The figure shows a profound effect of flexibility: the number of items that can be identified with a maximum allowable error probability of, in this case, 25%, is no longer a fixed factor 2.4 smaller than the dynamic range S/σ, and rises asymptotically to a value that is essentially determined by the variability $\hat{s}$ caused by the flexibility of the metric. This result is in good correspondence with a well-known experimentally established fact: when subjects are asked to identify stimuli that vary in one-dimensional attributes like loudness, brightness, or length, they are usually unable to successfully identify the stimuli with more than the famous "seven plus or minus two" stimulus categories (Miller 1956), even though the number of discriminable steps in these attributes is on the order of 100.

Within the framework of (partially) flexible metrics we have found a tentative explanation for this effect: for a rigid metric we have estimated that the number of identifiable items should be about 2.4 times smaller than the dynamic range S/σ, whereas the number of discriminable steps should be essentially equal to the dynamic range. Due to the flexibility of the metric, the additional variability $\hat{s}$ in the mapping from attribute strength x to scale value s essentially becomes the limiting factor for the number of identifiable items, whereas the dynamic range S/σ now becomes the lower limit for the number of discriminable steps (refer to what has been concluded earlier about the number of JNDs for partially flexible metrics). We therefore expect that the ratio of the number of discriminable steps to the number of identifiable items will be around 2.4 for those attributes for which the metric is rigid, with values exceeding 2.4 for those attributes whose internal metrics are increasingly more flexible.

6.7.4 Optimizing the degree of flexibility

Our aim here is to give an example of the solution to the flexibility optimization problem [Eq. (6.41)] for a simplified situation. To this end, we have assumed that the average distribution $\mu(x)$ is uniform between $x = 0$ and $x = 1$ and the momentary distribution $\mu(x, t)$ is Normal between $x = 0$ and $x = 1$ with mean 0.5 and spread 0.15. Furthermore, we have assumed that the scale range S is one, and we have assumed a dynamic range S/σ of 100. For the noise level σ, these choices result in a value of 0.01. Last, we have assumed that the number of items to be discriminated, N, is 100 (the maximum number of discriminable items for the chosen dynamic range of 100 and for a maximum allowable error probability of 25%), and that the number of items to be identified, N_i, is 10 (a number that for a partially flexible metric, a dynamic range of 100, and a maximum allowable error probability of 25% is close to the maximum number of identifiable items). Figure 6.3 shows a plot of discriminability D (dotted line) and identifiability I (dashed line) versus the degree of flexibility λ. The solid line represents total performance T when the value of ϵ in Eq. (6.41) is set to 0.5.

The plot indeed shows that discriminability rises monotonically with the degree of flexibility. It also shows that, for a degree of flexibility below ap-

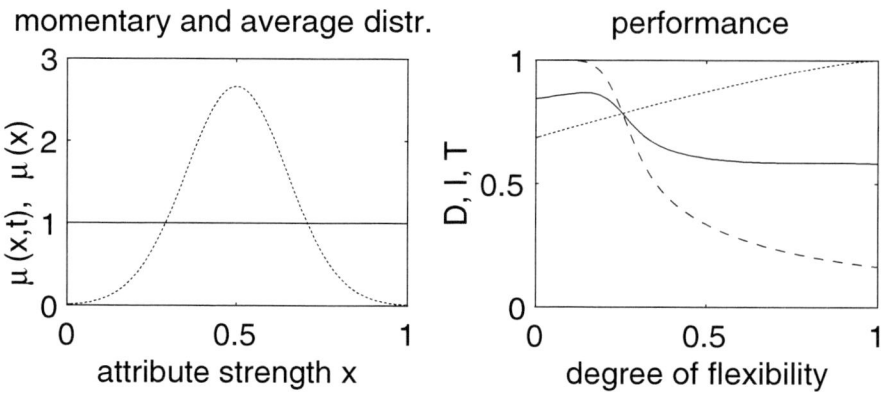

Figure 6.3: The influence of the degree of flexibility on discriminability D, identifiability I, and total performance T for a simplified case. The left panel shows the assumed momentary distribution $\mu(x,t)$ (dotted line) and average distribution $\mu(x)$ (solid line). The right panel shows calculated discriminability (dotted line), identifiability (dashed line) and total performance (solid line) versus the degree of flexibility. For this particular case, total performance is optimized for a degree of flexibility of 0.15. See text for further details about the assumed values for parameters such as dynamic range.

proximately 0.20, identifiability is relatively unaffected by increasing flexibility. However, for degrees of flexibility of about 0.25, identifiability decreases dramatically with increasing flexibility. Figure 6.3 shows that improved discriminability compensates for decreased identifiability up to a degree of flexibility of 0.15, where total performance T reaches its maximum of approximately 0.87. Therefore, for this specific case a relatively small degree of flexibility of 0.15 optimizes overall performance.

To conclude, the particular degree of flexibility that optimizes overall performance will in general depend on the relative importances of discriminability and identifiability, with increasing importance of discriminability resulting in a higher degree of flexibility and, conversely, increasing importance of identifiability, resulting in a lower degree of flexibility. Quantification of outside world attributes is an early stage in perception, and we therefore expect flexible metrics to be a property of the earliest stages in perception. Having observed this, it is unlikely that the degree of flexibility can be changed from one moment to the next according to what is required by the task at hand. Instead, it is more likely that the optimal choice for flexibility will be given by the successfulness of the interaction process during a longer period in time; for example, in terms of the number and severity of the errors made during the interaction.

An important parameter determining how discriminability and identifiability will vary for a given degree of flexibility is the statistical behavior of the momentary distribution in time. Luminance is a particularly interesting case in this respect. A typical range of luminance is about three orders of magnitude at any particular moment; however, the total range of luminance is estimated to be about 10 orders of magnitude (McCann 1988). For a rigid metric this entire range is somehow to be mapped onto the internal scale; therefore, even for a strongly compressive scale function, such as a logarithmic scale function, only about 30 percent of the scale would really be used at any particular point in time. For a flexible metric this situation is dramatically improved, with the lowest luminance in the scene mapped onto the lower end of the scale, and the highest luminance in the scale mapped onto the upper end of the scale. In this situation, scale values will roughly correspond to *relative* luminances, and therefore to surface reflection coefficients. Since surface reflection can be regarded as an invariant object property, the use of a flexible metric in this particular case actually helps to identify items using their measured luminances.

6.8 Application: black-and-white images of natural scenes

Our aim in this section is to compare discriminability, identifiability, and total performance predictions for four black-and-white images of natural scenes with experimentally obtained judgments of sharpness, visibility of detail, naturalness, and quality. Judgments of sharpness and visibility of detail were chosen since these attributes are in different ways related to discriminability. Judgments of naturalness, which in the experiment was defined as the degree to which the image is realistically reproduced, were chosen since this attribute is indirectly related to identifiability; the less realistically an image is reproduced, the less well the image can be matched with what is in memory, and therefore the less identifiable the items in the image will be. The aim of this comparison will not be to analyze the accuracy with which the predictions correspond to experimental results. Instead, the aim will be to check whether several nontrivial tendencies in the obtained experimental data, most notably the differences between naturalness judgments and quality judgments reported earlier in Chapter 3, can be adequately predicted.

We selected our images from a set of 77 digitized color images of natural scenes recorded on two Kodak Photo CDs. We produced black-and-white versions of these images by transforming the images from the RGB system space to the CIELUV color space (Hunt 1992) and setting the chromaticity coordinates u^* and v^* to zero. Finally, we manipulated the brightness contrast of the obtained black-and-white images by applying an s-shaped transformation on lightness L^*:

$$L^{*'} = \left(\frac{L^* - L^*_{\min}}{L^*_{\text{ave}} - L^*_{\min}} \right)^{\gamma} (L^*_{\text{ave}} - L^*_{\min}) + L^*_{\min}$$

$$(\text{for } L^*_{\min} \leq L^* \leq L^*_{\text{ave}})$$

$$L^{*'} = \left(\frac{L^*_{\max} - L^*}{L^*_{\max} - L^*_{\text{ave}}} \right)^{\gamma} (L^*_{\text{ave}} - L^*_{\max}) + L^*_{\max}$$

$$(\text{for } L^*_{\text{ave}} < L^* \leq L^*_{\max}), \tag{6.52}$$

where L^* represents the original lightness value, $L^{*'}$ the new value, and where $L^*_{\min}$, $L^*_{\max}$, and L^*_{ave} represent the minimum, maximum, and average lightness in the original image, respectively. The parameter γ in Eq. (6.52) is specified in terms of a gain factor g as

$$\gamma = 10^g. \tag{6.53}$$

Applying the above s-shaped transformation will decrease the brightness contrast for negative values of g and increase the brightness contrast of the image for positive values of g, while the average lightness of the image remains at approximately the original level. Nine versions of each scene were used, with gain factor values of -0.60, -0.45, -0.30, -0.15, 0, 0.15, 0.30, 0.45, and 0.60. Subjects, eight in number, were in separate sessions instructed to judge sharpness, visibility of detail, naturalness, and quality of the resulting images. The images were shown in random order and with three replications on a PAL-compliant CRT. Subjects were instructed to use an 11-point numerical scale ranging from 0 ("bad") to 10 ("excellent") for their judgments. The resulting judgments were z-scored and averaged over subjects and replications.

For our predictions we assumed the following. First, for the momentary distribution $\mu(x, t)$ we used the luminance distribution of the image under consideration, and for the average distribution $\mu(x)$ we used the average luminance distribution of the entire set of 77 images. We again assumed a scale range of one, a dynamic range of 100, and 100 items to be discriminated. Furthermore, we assumed equal weighting of discriminability and identifiability for overall importance ($\epsilon = 0.5$) and, in correspondence with what was found for the simplified case in the previous section, a relatively small degree of flexibility, $\lambda = 0.25$. This particular choice for the degree of flexibility was also found to best describe experimental results in the color domain (see Chapter 5). Last, we checked for our set of 77 images whether the deviation $\Delta s(x, t)$ [Eq. (6.44)] could indeed be adequately described by a Normal probability density function. This proved to be the case, and the spread $\hat{s}$ [Eq. (6.50)] that we found for the set of images was approximately 0.128λ. Using Eq. (6.36), we arrived at a maximum number of identifiable items of approximately 13 for the above choices of $\lambda = 0.25$ and $P_i = 25\%$ (see also Table 6.1).

Figure 6.4 shows judgments of visibility of detail (dotted lines) and sharpness (dashed lines) together with z-scored discriminability predictions (solid lines). For images 2 and 3 discriminability predictions closely correspond to sharpness judgments, for image 1 predictions are roughly midways the judgments of sharpness and visibility of detail, and for image 4 predictions closely correspond to judgments of visibility of detail. The difference between sharpness and visibility of detail for values of g larger than zero is probably due to two separate effects of the s-shaped transform. For small positive values of g both sharpness and visibility of detail are increased, due to increased contrast. However, for large positive values of g visibility of detail is decreased, due to the presence of increasingly larger areas in the manipulated images where all detail is lost due to clipping to

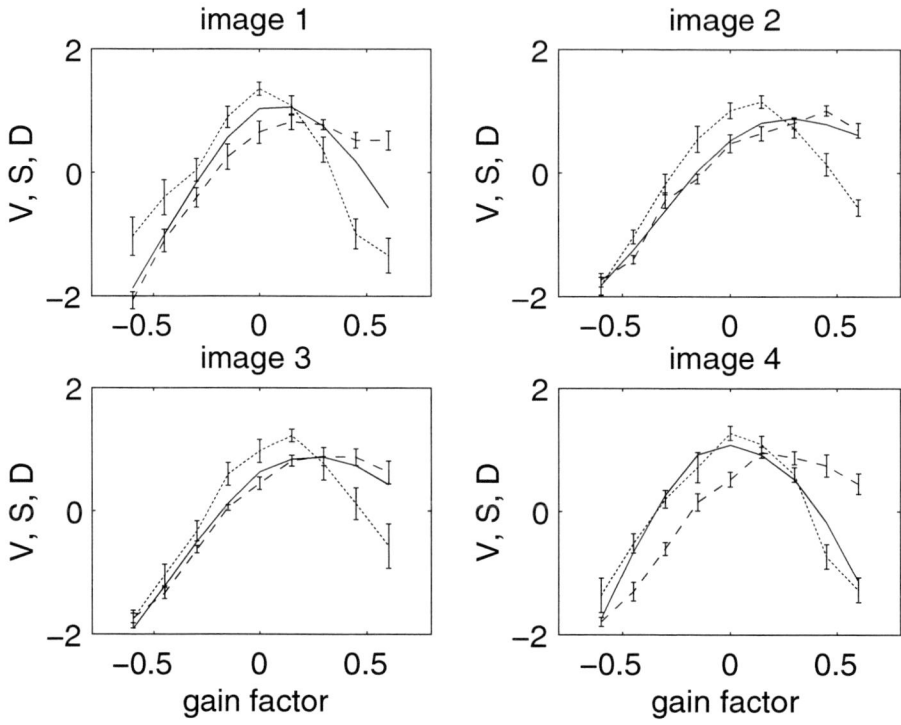

Figure 6.4: Experimentally obtained judgments for visibility of detail V (dotted lines) and sharpness S (dashed lines), versus the gain factor g of the s-shaped transform. The error bars denote twice the standard error in the mean. The solid lines represent the predictions for discriminability D. Both judgments and predictions have been z-scored to facilitate comparison.

Table 6.1: Calculated number of identifiable items for the luminance distributions of the set of 77 images, assuming $S = 1.00$ and $\sigma = 0.01$ (dynamic range $S/\sigma = 100$). The number of identifiable items was calculated for degrees of flexibility λ of 0.00, 0.25, 0.50, 0.75, and 1.00, and for a maximum allowable error probability P_i of 25% and 5%, For the predictions in this section we have chosen $P_i = 25\%$ and $\lambda = 0.25$, hence $N_i \approx 13$.

$P_i = 25\%$						$P_i = 5\%$				
λ	0.00	0.25	0.50	0.75	1.00	λ	0.00	0.25	0.50	0.75	1.00
N_i	43.5	12.9	6.7	4.5	3.4	N_i	25.5	7.6	3.9	2.6	2.0

either black or white. Sharpness, which is closely related to contrast, is not influenced much by the existence of such areas. It is not exactly clear to us why predictions correspond more closely to sharpness for some images and to visibility of detail for other images, and in the remainder of this discussion we have therefore chosen to consider the average of the judgments for sharpness and visibility of detail as an adequate correlate for discriminability.

Figure 6.5 shows naturalness judgments (dotted lines) and identifiability predictions (solid lines). Agreement between predictions and judgments is quite good for image 4, and somewhat less good for images 1, 2 and 3. The naturalness judgments for these three images show some asymmetry, with naturalness being judged best for small positive values of g. The position of the maximum of the naturalness judgments for these images nevertheless corresponds nicely with the position of the maximum of the identifiability predictions. The most important difference is the increased asymmetry of the identifiability predictions with respect to the naturalness judgments.

Figure 6.6 shows quality judgments (dotted lines) and predictions for total performance (solid lines). Here, agreement between predictions and judgments is quite good for all images. The only important difference is the overestimation of the predictions for total performance with respect to quality judgments for image 2 for large positive values of g.

Figure 6.7 shows the average of the judgments for sharpness and visibility of detail (dotted lines), naturalness judgments (dashed lines) and quality

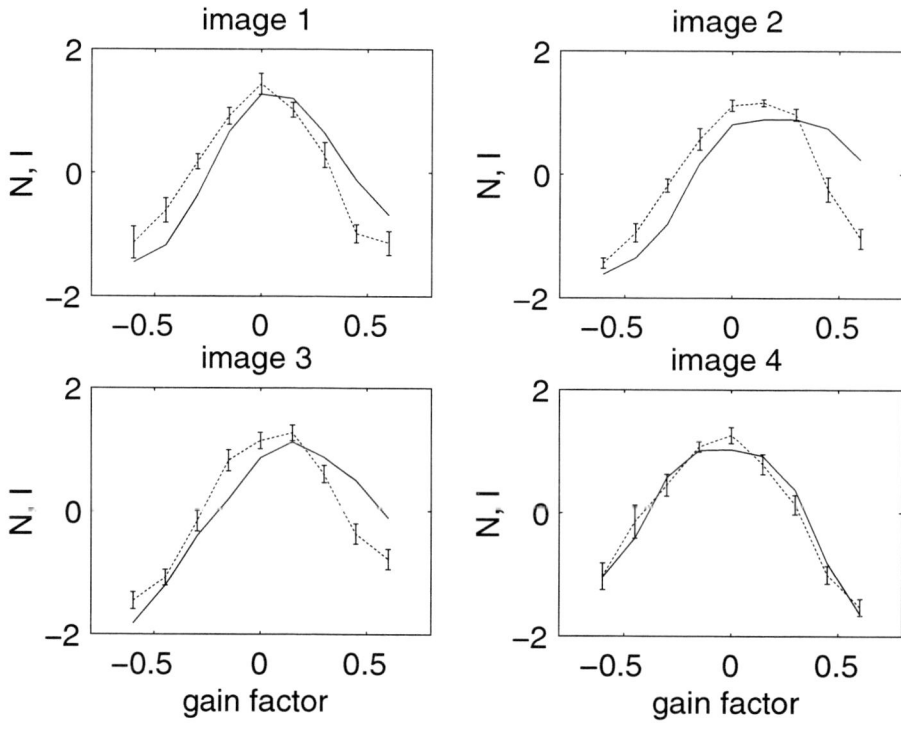

Figure 6.5: Experimentally obtained judgments for naturalness N (dotted lines) and predictions for identifiability I (solid lines), versus the gain factor g of the s-shaped transform. The error bars denote twice the standard error in the mean. Judgments and predictions have been z-scored to facilitate comparison.

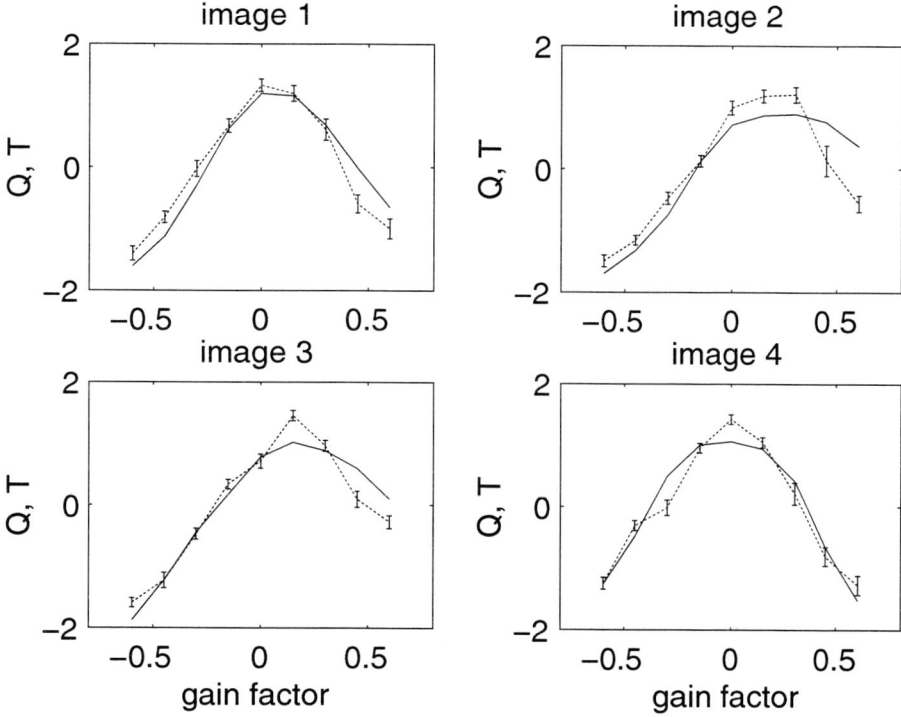

Figure 6.6: Experimentally obtained judgments for quality Q (dotted lines) and predictions for total performance T (solid lines), versus the gain factor g of the s-shaped transform. The error bars denote twice the standard error in the mean. Judgments and predictions have been z-scored to facilitate comparison.

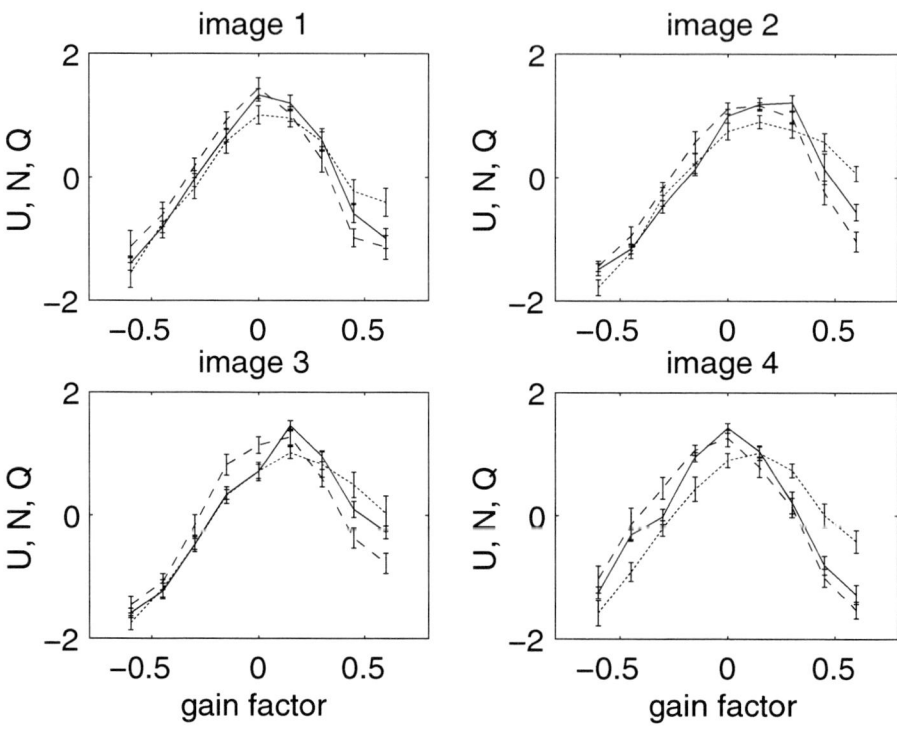

Figure 6.7: The average of the experimentally obtained judgments for visibility of detail and sharpness, U (dotted lines), experimentally obtained judgments of naturalness N (dashed lines), and experimentally obtained judgments of quality Q (solid lines) versus the gain factor g of the s-shaped transform. The error bars denote twice the standard error in the mean. All judgments have been z-scored to facilitate comparison.

judgments (solid lines). The figure shows that quality judgments lie approximately halfway naturalness judgments and the average of the judgments for sharpness and visibility of detail. This result strongly supports the central idea of this chapter that quality is determined by both discriminability and identifiability.

To conclude, Fig. 6.8 shows predictions for discriminability (dotted lines), identifiability (dashed lines), and overall performance (solid lines). Compare the predictions in this figure with the experimental results in Fig. 6.7. Although smaller in size, the differences between discriminability and identifiability correspond nicely to the differences between the average of sharpness and visibility of detail and naturalness. Especially the small but systematic differences between naturalness and quality, with quality being higher than naturalness for large positive values of g (all images) and quality being lower than naturalness for negative values of g (most clearly in image 4), are correctly reflected in the predictions of identifiability and total performance. Therefore, although there are significant differences between experimental results and predictions, the validity of the concept we have developed here is supported by the correct prediction of several nontrivial characteristics of the obtained experimental results.

6.9 Conclusions

We have presented a concept for image quality that is based on answering the questions: What are images? What are images used for? and What are the requirements which the use of images imposes on them? We have argued that images are the carriers of visual information, and that they are used as input to the perception stage of interaction. Furthermore, we have argued that the aim of perception is to measure and internally quantify attributes of items in the outside world in order to discriminate and identify these items. Therefore, the requirements imposed on an image are that, first, the items in the image should be discriminable and, second, the items in the image should be identifiable.

Since the internal quantification of outside world attributes is of such importance for a successful interaction, we have argued that the metrics used for this quantification should be of the type proposed in Chapter 4, that is,

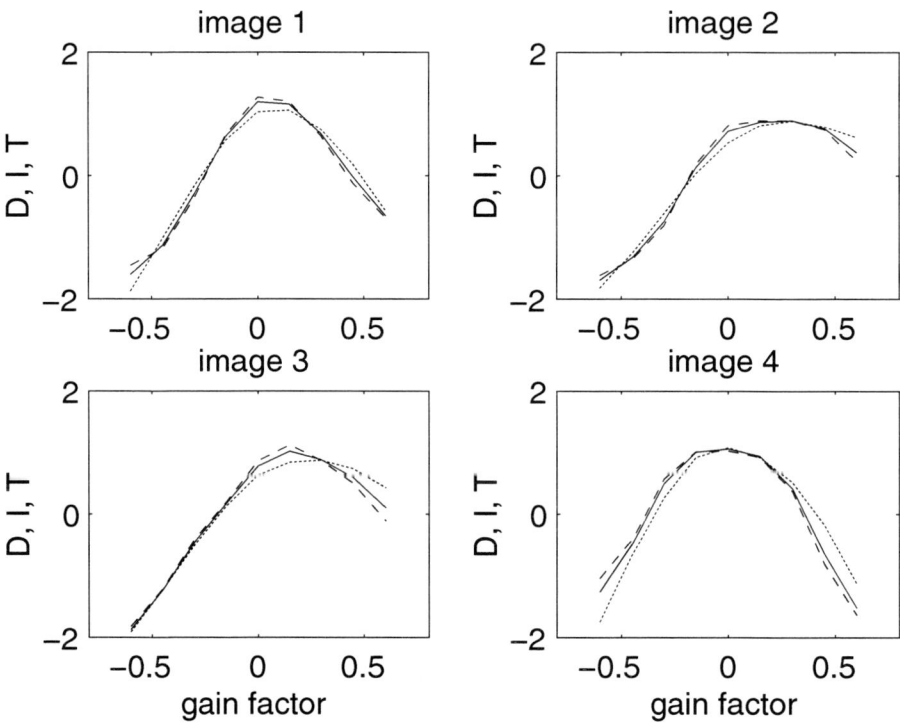

Figure 6.8: Predictions for discriminability D (dotted lines), identifiability I (dashed lines) and total performance T (solid lines) versus the gain factor g of the s-shaped transform.

partially flexible and optimized with respect to discriminability and identifiability. We have derived expressions for such metrics, together with measures for discriminability, identifiability, and total performance. For the case of manipulated black-and-white reproductions of natural scenes, we have compared predictions made using these algorithms with experimentally obtained judgments of human subjects. Although there are significant differences between predictions and judgments, correspondence between predictions and judgments is quite good. Moreover, several important and nontrivial characteristics of the obtained judgments, such as the small but systematic difference between quality and naturalness judgments, are correctly predicted.

Chapter 7

Epilogue

In this book a new concept for image quality has been presented. The concept is based on a top-down analysis of visuo-cognitive processing, in particular of the role of visuo-cognitive processing within the interaction process. Starting with a fundamental definition of quality in terms of the degree to which imposed requirements are satisfied, we have obtained a description of image quality that takes into account explicitly the fact that images are input to the vision stage of the interaction process. The result is a definition for image quality in terms of the adequacy of the image to serve as an input for the interaction process.

As stated in the conclusion of Chapter 3, the concept we have presented here is formulated basically independently of modality, which opens the possibility to apply this concept to, for example, sound or speech quality, or to generalize it to the quality of multimodal information presentation. An interesting possible extension mentioned in the conclusion of Chapter 3 is a generalization toward a concept for the quality of user-system interaction. The topic of this final chapter will be a short discussion of two (philosophical) issues related to the development of such a concept.

To start with the first issue, Newell & Simon (1972) and Newell (1990) have argued convincingly that various aspects of human behavior, such as reasoning, search, and problem solving, and therefore interaction in general, can be regarded as information-processing tasks. There is an important philosophical consequence to this position, namely that humans may be regarded as information-processing systems. Marr (1982) has argued that, in order to fully understand information-processing systems, these systems should be studied at the levels of *semantics*, *algorithms*, and *implementation*. Typical questions asked at these three levels are: what is the system doing, and why is it doing what it is doing (semantics), how is it doing what it is doing (algorithms), and what is it using to do what it is doing (implementation).

The second issue we wish to consider here is the *structure* of information-processing tasks, which seems to be strongly *hierarchical*. This is probably due to the fact that most information-processing tasks are performed by means of sequences of procedures, where each individual procedure often is an information-processing task in its own right. Hence, information-processing tasks can be subdivided into smaller tasks (subtasks) and are themselves part of larger tasks (supertasks). An important consequence of this hierarchical structure is that the goal of an information-processing task is in fact defined *one level upward* in the hierarchy of information-processing tasks being performed.

The above two observations have a profound influence on the meaning of the concept quality within user-system interaction. Two consequences are particularly interesting here, and we will discuss them in order. The first is the fact that the concept of quality can be defined hierarchically across the three levels of semantics, algorithm, and implementation. The second is the fact that, at the semantic level, quality can only be defined meaningfully when the hierarchical structure of information processing is taken explicitly into account.

If we consider the semantic level, quality can be defined (as before in this book) in terms of the adequacy of the information-processing task for reaching the desired goal. Suitable criteria at the semantic level may for example be: has the desired goal been successfully accomplished (effectivity) and at what effort (efficiency). At the algorithmic level, quality refers to the adequacy of the algorithms that are performing the information-processing task, with criteria such as the ability to work on distorted input (robustness) or applicability to a wide range of different inputs (flexibility). Finally, at the level of implementation, quality refers to the adequacy of the hardware used to implement the information-processing algorithms. Typical criteria at this level are physical limits to the precision with which signals can be represented (noise), or representable signal strength (range).

The level of implementation is the level where quality is usually defined. Think, for example, of quality definitions for audio-visual equipment, where quality is defined in terms of limits to the precision and range with which signals can be represented. In this book we have successfully developed an alternative concept for image quality at the semantic level. We have made this choice for one very important reason: quality at the level of implementation, or at the algorithmic level, is a *necessary but insufficient condition* for quality at the semantic level. In other words: although quality at the algorithmic and implementation level is important, and although quality at the semantic level depends upon quality at these two levels, it is quality at the semantic level that ultimately determines how well humans are able to *use* the image. This will not be different for a possible future concept for the quality of user-system interaction, where the user interface takes the role the image has in the present concept.

As stated earlier, the nature of information-processing tasks, and hence of user-system interaction, is strongly hierarchical. We will start at what is perhaps the lowest distinguishable hierarchical level in user-system interaction: the level of the individual user-system interaction cycle. Furthermore, we will assume that the goal of user-system interaction at this lowest hierarchical level is to manipulate the system from its current state to some desired state, as specified one level upward in the hierarchy. Figure 7.1 shows the anatomy of a single user-system interaction cycle.

The figure shows two flows of information between the user and the system, both of which are mediated by the user interface. The information flow going from the system to the user is typically encoded visually, for example by means of images displayed on a screen; and is usually about the current status of the system. The information flow going in the opposite direction, from the user to the system, is usually encoded motorically, for example by means of mouse movements and mouse clicks; and often contains instructions from the user to the system. The user part of the interaction cycle in Fig. 7.1 is again divided into the stages perception, cognition, and action. Within this division, the aim of the perception stage is to "decode" the information flow from the system to the user. In the cognition stage, this decoded information is used to (re)formulate the strategy for achieving the desired system state and deriving a procedure of instructions to the system. The aim of the action stage is then to "encode" these instructions into a for-

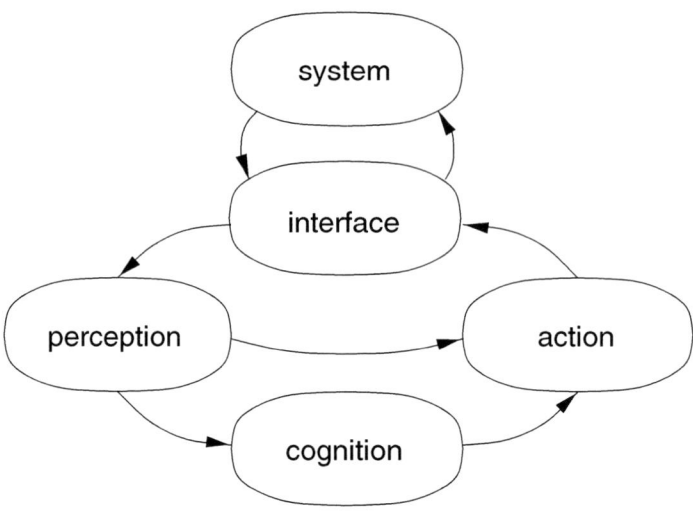

Figure 7.1: Anatomy of the user-system interaction cycle.

mat that can be interpreted by the interface and subsequently mediated to the system.

We have already observed that, at the semantic level, quality can only be meaningfully defined when the hierarchical structure of information processing tasks is taken into account. We are now at a point where we are able to explain why. To this end, consider that at the level of implementation, quality may be defined in terms of the adequacy of the user-interface hardware as a means to mediate system status information to the user and user instructions to the system. Similarly, quality may be defined at the level of algorithms in terms of the adequacy of the formulated set of instructions to perform the task of manipulating the system from its present state to some new state, using the available user interface. Finally, at the semantic level, quality may be defined in terms of the adequacy of the interaction strategy with respect to *successfully*, that is, effectively and efficiently, achieving the *desired* final system state. It is exactly for this that we need to know what the role of the interaction task is within its interaction supertask one level upward in the hierarchy.

To conclude, the development of a concept for user-system interaction quality will require an in-depth analysis of the structure of the interaction process, as well as a careful analysis of the definitions for quality at the levels of semantics, algorithms, and implementation. A possible means to achieve this may be the definition of relatively well defined, autonomous information-processing tasks at higher levels in the information-processing task hierarchy, such as the search task, similar to the discrimination and identification tasks we have defined earlier for visuo-cognitive processing. The knowledge thus obtained can be used to formulate requirements that should be imposed on user interfaces in order to ensure a successful interaction, and subsequently to develop measures for quantifying the degree to which these requirements are satisfied. The availability of such measures will be an extremely important step toward the design of high-quality user interfaces.

Bibliography

Andrews, D. (1964). Error-correcting perceptual mechanisms, *Quarterly Journal of Experimental Psychology* **16**: 104–115.

Ballard, D. & Brown, C. (1982). *Computer Vision*, Prentice-Hall, Englewood Cliffs.

Barlow, H. (1961). Possible Principles Underlying the Transformation of Sensory Messages, *in* W. Rosenblith (ed.), *Sensory Communication*, Wiley, London, pp. 217–234.

Barrow, H. & Tenenbaum, J. (1986). Computational Approaches to Vision, *in* K. Boff, L. Kaufman & J. Thomas (eds), *Handbook of Perception and Human Performance*, Wiley-Interscience, Chichester, pp. 38.1–38.70.

Barten, P. (1990). Evaluation of subjective image quality with the square-root integral method, *Journal of the Optical Society of America A* 7(10): 2024–2031.

Blommaert, F. (1995). Visual metrics, flexibility, and usefulness, *Perception* **24**(supp): 93.

Boschman, M. & Roufs, J. (1997). Text quality metrics for visual display units: II. An experimental survey, *Displays* **18**: 45–64.

Bruce, V. & Green, P. (1985). *Visual Perception: Physiology, Psychology and Ecology*, Erlbaum, London.

Cohn, T. & Lasley, D. (1986). Visual sensitivity, *Annual Review of Psychology* **37**: 495–521.

Cosman, P., Gray, R. & Olshen, R. (1994). Evaluating quality of compressed medical images: SNR, subjective rating, and diagnostic accuracy, *Proceedings of the IEEE* **82**(6): 919–932.

Daly, S. (1993). The Visible Differences Predictor: an Algorithm for the Assessment of Image Fidelity, *in* A. Watson (ed.), *Digital Images and Human Vision*, MIT Press, Cambridge MA, pp. 179–206.

de Ridder, H. (1992). Minkowski-metrics as a combination rule for digital-image-coding impairments, *Proceedings of SPIE: Human Vision, Visual Processing, and Digital Display III, San Jose, 10–13 February 1992*, Vol. 1666, SPIE, Bellingham, pp. 16–26.

de Ridder, H. (1996). Naturalness and image quality: saturation and lightness variation in color images of natural scenes, *Journal of Imaging Science and Technology* **40**(6): 487–493.

de Ridder, H., Fedorovskaya, E. & Blommaert, F. (1993). Naturalness and image quality: chroma variations in colour images of natural scenes, *IPO Annual Progress Report* **28**: 89–95.

Eckert, M. & Bradley, A. (1998). Perceptual quality metrics applied to still image compression, *Signal Processing* **70**: 177–200.

Eimer, M. (1990). Representational content and computation in the human visual system, *Psychological Research* **52**: 238–242.

Falmagne, J.-C. (1985). *Elements of Psychophysical Theory*, Clarendon Press, Oxford.

Fechner, G. (1860). *Elemente der Psychophysik*, Breitkopf, Leipzig.

Fedorovskaya, E., de Ridder, H. & Blommaert, F. (1997). Chroma variations and perceived quality of color images of natural scenes, *Color: Research and Application* **22**(2): 96–110.

Gescheider, G. (1988). Psychophysical scaling, *Annual Review of Psychology* **39**: 169–200.

Gibson, J. (1950). *The Perception of the Visual World*, Allen and Unwin, London.

Gibson, J. (1966). *The Senses Considered as Perceptual Systems*, Houghton Mifflin, London.

Gibson, J. (1979). *The Ecological Approach to Visual Perception*, Houghton Mifflin, London.

Hermiston, K. & Booth, D. (1999). NIIRS and objective image quality measures, *Proceedings of CAIP 99: Computer Analysis of Images and Patterns, Ljubljana, Slovenia, 1–3 September 1999*, Springer, Berlin, pp. 385–394.

Hunt, R. (1992). *Measuring Colour*, second edn, Ellis Horwood, Chichester.

Hurlbert, A. (1986). Formal connections between lightness algorithms, *Journal of the Optical Society of America A* **3**(10): 1684–1693.

Irving, J. & Mullineux, N. (1959). Mathematics in Physics and Engineering, *in* H. Massey & K. Bruecker (eds), *Pure and Applied Physics*, Vol. 6, Academic Press, New York.

Janssen, T. & Blommaert, F. (1997). Image quality semantics, *Journal of Imaging Science and Technology* **41**(5): 555–560.

Janssen, T. & Blommaert, F. (2000a). A computational approach to image quality, *Displays* **21**(4): 129–142.

Janssen, T. & Blommaert, F. (2000b). Predicting the usefulness and naturalness of colour reproductions, *Journal of Imaging Science and Technology* **44**(2): 93–104.

Janssen, T. & Blommaert, F. (2000c). Visual metrics: discriminative power through flexibility, *Perception* **29**(8): 965–980.

Kayargadde, V. & Martens, J.-B. (1996a). Perceptual characterization of images degraded by blur and noise: experiments, *Journal of the Optical Society of America A* **13**(6): 1166–1177.

Kayargadde, V. & Martens, J.-B. (1996b). Perceptual characterization of images degraded by blur and noise: model, *Journal of the Optical Society of America A* **13**(6): 1178–1188.

Laming, D. (1986). *Sensory Analysis*, Academic press, London.

Leachtenauer, J. (2000). Comparison of video compression evaluation metrics for military applications, *Proceedings of SPIE: Human Vision and Electronic Imaging V, San Jose, 24–27 January 2000*, Vol. 3959, SPIE, Bellingham, pp. 88–98.

Lubin, J. (1995). A Visual Discrimination Model for Imaging System Design and Evaluation, *in* E. Peli (ed.), *Vision Models for Target Detection and Recognition*, World Scientific, London, pp. 245–283.

Luce, R. & Narens, L. (1987). Measurement scales on the continuum, *Science* **236**(4808): 1527–1532.

Marr, D. (1982). *Vision: a Computational Investigation into the Human Representation and Processing of Visual Information*, Freeman, San Francisco.

Martens, J.-B. & Meesters, L. (1999). The role of image dissimilarity in image quality models, *Proceedings of SPIE: Human Vision and Electronic Imaging IV, San Jose, 25–28 January 1999*, Vol. 3644, SPIE, Bellingham, pp. 258–269.

McCann, J. (1988). Calculated color sensations applied to image reproduction, *Proceedings of SPIE: Image Processing, Analysis, Measurement, and Quality, Los Angeles, 13–15 January 1988*, Vol. 901, SPIE, Bellingham, pp. 205–214.

Miller, G. (1956). The magical number seven, plus or minus two: some limits on our capacity for processing information, *Psychological Review* **63**(2): 81–97.

Newell, A. (1990). *Unified Theories of Cognition*, Harvard University Press, Harvard.

Newell, A. & Simon, H. (1972). *Human Problem Solving*, Prentice-Hall, Englewood Cliffs.

Nijenhuis, M. & Blommaert, F. (1997). Perceptual error measure for sampled and interpolated images, *Journal of Imaging Science and Technology* **41**(3): 249–258.

Poggio, T. & Koch, C. (1985). Ill-posed problems in early vision: from computational theory to analogue networks, *Proceedings of the Royal Society of London B* **226**(1244): 303–323.

Poggio, T., Torre, V. & Koch, C. (1990). Computational Vision and Regularization Theory, *in* S. Ullman & W. Richards (eds), *Image Understanding 1989*, Ablex, Norwood, pp. 1–18.

Riesz, R. (1933). The relation between loudness and the minimum perceptible increment of intensity, *Journal of the Acoustical Society of America* **5**: 211–216.

Roufs, J. (1993). Perceptual image quality: concept and measurement, *Philips Journal of Research* **47**: 35–62.

Roufs, J. & Boschman, M. (1997). Text quality metrics for visual display units: I. Methodological aspects, *Displays* **18**: 37–43.

Stevens, S. (1957). On the psychophysical law, *Psychological Review* **64**: 153–181.

Takasaki, H. (1966). Lightness change of grays induced by change in reflectance of gray background, *Journal of the Optical Society of America* **56**(4): 504–509.

Thomas, J. (1969). *An Introduction to Statistical Communication Theory*, Wiley, London.

Wallach, H. (1948). Brightness constancy and the nature of achromatic colors, *Journal of Experimental Psychology* **38**(3): 310–324.

Watt, R. (1989). *Visual Processing: Computational, Psychophysical and Cognitive Research*, Erlbaum, Hove.

Watt, R. (1991). *Understanding Vision*, Academic Press, London.

Weber, E. (1846). Der Tastsinn und das Gemeingefühl, *Handwörterbuch der Physiologie* **3**: 481–588.

Whittle, P. (1994a). Contrast Brightness and Ordinary Seeing, *in* A. Gilchrist (ed.), *Lightness, Brightness, and Transparency*, Erlbaum, Hillsdale, pp. 35–110.

Whittle, P. (1994b). The Psychophysics of Contrast Brightness, *in* A. Gilchrist (ed.), *Lightness, Brightness, and Transparency*, Erlbaum, Hillsdale, pp. 111–157.

Winkler, S. (1999). Issues in vision modeling for perceptual video quality assessment, *Signal Processing* **78**: 231–252.

Yendrikhovskij, S., Blommaert, F. & de Ridder, H. (1999a). Colour reproduction and the naturalness constraint, *Color: Research and Application* **24**(1): 52–67.

Yendrikhovskij, S., Blommaert, F. & de Ridder, H. (1999b). Representation of memory prototype for an object colour, *Color: Research and Application* **24**(6): 393–410.

Zetzsche, C. & Hauske, G. (1989). Multiple channel model for the prediction of subjective image quality, *Proceedings of SPIE: Human Vision, Visual Processing, and Digital Display, Los Angeles, 18–20 January 1989*, Vol. 1077, SPIE, Bellingham, pp. 209–216.

Zhang, X., Setiawan, E. & Wandell, B. (1997). Image distortion maps, *Proceedings of the Fifth Color Imaging Conference: Color Science, Systems, and Applications, Scottsdale, 17–20 November 1997*, Society for Imaging Science and Technology, Springfield, pp. 120–125.

Zhang, X. & Wandell, B. (1998). Color image fidelity metrics evaluated using image distortion maps, *Signal Processing* **70**: 201–214.

Sample stimuli

The following pages show three sets of sample stimuli similar to those used in the experiments described in Chapter 3. The first set (page 141) shows the effect of scaling CIE chroma, C_{uv}^*. The chroma scale factors used are: 0.50 (upper left), 0.71 (upper right), 1.00 (center), 1.41 (lower left), and 2.00 (lower right). The second set (page 142) shows the effect of varying the color temperature of the reference white, T_c. The color temperatures used are: 4,000 K (upper left), 5,000 K (upper right), 6,500 K (center), 10,000 K (lower left), and 25,000 K (lower right). The third and last set (page 143) shows the effect of applying the s-shaped transformation on CIE lightness, L^*. The values of γ used are: 0.25 (upper left), 0.50 (upper right), 1.00 (center), 2.00 (lower left), and 4.00 (lower right). The corresponding values for the gain factor g are -0.6, -0.3, 0.0, 0.3, and 0.6, respectively.

142

Author Biography

Ruud Janssen studied Electrical Engineering at the Eindhoven University of Technology (TUE), in Eindhoven, The Netherlands. During this study he specialized in signal processing and medical electrical engineering. After finishing this study, he did his Ph.D. research in image quality and color reproduction at the Institute for Perception Research (IPO), also in Eindhoven, The Netherlands. For this research he was rewarded a Cum Laude doctoral degree in 1999. He now works in the Research & Development department of Océ-Technologies, in Venlo, The Netherlands.